AF509924

Performance of Mechanical Properties of Ultrahigh-Strength Ferrous Steels Related to Strain-Induced Transformation

Performance of Mechanical Properties of Ultrahigh-Strength Ferrous Steels Related to Strain-Induced Transformation

Editor

Koh-ichi Sugimoto

MDPI • Basel • Beijing • Wuhan • Barcelona • Belgrade • Manchester • Tokyo • Cluj • Tianjin

Editor
Koh-ichi Sugimoto
Shinshu University
Japan

Editorial Office
MDPI
St. Alban-Anlage 66
4052 Basel, Switzerland

This is a reprint of articles from the Special Issue published online in the open access journal *Metals* (ISSN 2075-4701) (available at: https://www.mdpi.com/si/metals/perform_mech_prop_ultrah_strength_ferr_steel_relat_strain_induc_transform).

For citation purposes, cite each article independently as indicated on the article page online and as indicated below:

LastName, A.A.; LastName, B.B.; LastName, C.C. Article Title. *Journal Name* **Year**, *Article Number*, Page Range.

ISBN 978-3-03943-428-2 (Hbk)
ISBN 978-3-03943-429-9 (PDF)

Contents

About the Editor

Koh-ichi Sugimoto, 1973 Bachelor of Engineering (Shinshu University), 1975 Master of Engineering (Shinshu University), 1985 Doctor of Engineering (Tokyo Metropolitan University), 1975–1985 Assistant Professor (Tokyo Metropolitan University), 1986 Assistant Manager (Daido Steel Co. Ltd.), 1987–1997 Associate Professor (Shinshu University), 1998–2015 Professor (Shinshu University), 2016 Professor Emeritus (Shinshu University).

Preface to "Performance of Mechanical Properties of Ultrahigh-Strength Ferrous Steels Related to Strain-Induced Transformation"

Ultrahigh-strength ferrous steels, related to the strain-induced martensite transformation (or transformation-induced plasticity: TRIP) of metastable retained austenite, such as TRIP-aided bainite/martensite steels, quenching and partitioning steels, nanostructured bainitic steels (or carbide free bainitic steels) and medium manganese steels, are currently receiving a great deal of attention from both academic and industry sectors, due to their excellent formability and mechanical properties. Many researchers are interested in the microstructure, retained austenite characteristics and tensile properties of ferrous steels, which enhance the strength-ductility balance. To further apply the ferrous steels to a wide range of automotive components and parts, a detailed understanding of the performance of the mechanical properties, such as impact toughness, fatigue strength, delayed fracture strength and wear resistance after heat-treatment, thermo-mechanical process, plastic working (including hot-stamping, hot-forging), welding, surface treatment, etc., will be a great help to steel engineers in the future. This book aims to address the performance of the mechanical properties of the ultrahigh-strength ferrous steels, as well as their strain-induced transformation behavior and the deformation mechanism.

Koh-ichi Sugimoto
Editor

Editorial

Performance of Mechanical Properties of Ultrahigh-Strength Ferrous Steels Related to Strain-Induced Transformation

Koh-ichi Sugimoto

Department of Mechanical Systems Engineering, School of Science and Technology, Shinshu University, Matsumoto 390-0802, Japan; sugimot@shinshu-u.ac.jp

Received: 29 June 2020; Accepted: 30 June 2020; Published: 1 July 2020

1. Introduction

Ultrahigh-strength ferrous steels, related to the strain-induced martensite transformation (or transformation-induced plasticity: TRIP) of metastable retained austenite, such as TRIP-aided bainite/martensite steels, quenching and partitioning steels, nanostructured bainitic steels (or carbide free bainitic steels) and medium manganese steels, are currently receiving a great deal of attention from both academic and industry sectors, due to their excellent formability and mechanical properties. Many researchers are interested in the microstructure, retained austenite characteristics and tensile properties of the ferrous steels, which enhance the strength-ductility balance. To further apply the ferrous steels to a wide range of automotive components and parts, a detailed understanding of the performance of the mechanical properties, such as impact toughness, fatigue strength, delayed fracture strength and wear resistance after heat-treatment, thermo-mechanical process, plastic working (including hot-stamping, hot-forging), welding, surface treatment, etc., will be a great help to steel engineers in the future.

This Special Issue aims to address the performance of the mechanical properties of the ultrahigh-strength ferrous steels, as well as their strain-induced transformation behavior and the deformation mechanism.

2. Contributions

One review and nine research articles have been published in this Special Issue of Metals. The subjects of this Issue are roughly divided into four categories: (1) strain-induced transformation behavior [1–3], (2) deformation mechanism [4,5], (3) formability [6,7] and (4) mechanical properties [8–10] of the ultrahigh-strength ferrous steels. The abovementioned (1) and (2) give important information for improving (3) and (4).

First, fundamental studies on the strain-induced transformation behavior of metastable retained austenite were conducted. Kaar et al. [1] studied the influence of quenching and partitioning processes on the transformation kinetics in a lean medium manganese TRIP steel. The paper by Grajcar et al. [2] dealt with the effect of the retained austenite size on the strain-induced transformation behavior during tensile deformation of hot-rolled multiphase steel. Hossain et al. [3] investigated the mechanical stability of retained austenite in high carbon steel, under compressive stress at different strain rates.

Second, the deformation mechanism of the ferrous steels was investigated using various microscopes. Tang et al. [4] reported the effect of the deformation temperature on the deformation mechanism of cold-rolled, low carbon, high manganese austenite/ferrite steel. Grzegorczyl et al. [5] reported the Portevin-Le Chatelier effect at 20 to 140 °C, in a hot-rolled, medium Mn bainite/martensite steel.

Third, the formability of the ferrous steels was related to the microstructure and the retained austenite characteristics. Kaar et al. [6] studied the structure-ductility relationship in cold-rolled medium manganese bainite/martensite steels subjected to quenching and partitioning processes. Yang et al. [7] investigated the springback behavior during the bending process for a quenching and partitioning steel.

Finally, mechanical properties, such as the impact toughness, fatigue strength and delayed fracture strength, of the ultrahigh-strength steels were investigated, and the fracture mechanisms were proposed. Sugimoto et al. [8] reported that chromium-molybdenum-bearing, hot-forged, medium carbon, TRIP-aided bainitic ferrite steels achieve excellent impact toughness. To apply the TRIP-aided martensitic steels to precision gears, Sugimoto et al. [9] reviewed the effects of the heat treatment and case-hardening processes on the fatigue strength, as well as hardness and residual stress in the surface hardening layer. Hojo et al. [10] studied the effects of alloying elements on the delayed fracture strength of TRIP-aided martensitic steels.

3. Conclusions and Outlook

A variety of topics concerning ultrahigh-strength ferrous steels were collected in this Special Issue. At present, most of the ferrous steels are applied to cold sheet parts. However, they may be used as the materials of hot-forged parts in the future, because of the excellent performance of the mechanical properties. I hope that many researchers will have an interest in the applications of the ferrous steels in the hot-forging of parts.

As a Guest Editor, I would like to thank all the authors for their academic contributions to this Special Issue. I would also like to give special thanks to all staff at the Metals Editorial Office.

Conflicts of Interest: The author declares no conflicts of interest.

References

1. Kaar, S.; Schneider, R.; Krizan, D.; Béal, C.; Sommitsch, C. Influence of the quenching and partitioning process on the transformation kinetics and hardness in a lean medium manganese TRIP steel. *Metals* **2019**, *9*, 353. [CrossRef]
2. Grajcar, A.; Kozłowska, A.; Radwański, K.; Skowronek, A. Quantitative analysis of microstructure evolution in hot-rolled multiphase steel subjected to interrupted tensile test. *Metals* **2019**, *9*, 1304. [CrossRef]
3. Hossain, R.; Pahlevani, F.; Sahajwalla, V. Evolution of microstructure and hardness of high carbon steel under different compressive strain rates. *Metals* **2018**, *9*, 580. [CrossRef]
4. Tang, Z.; Huang, J.; Ding, H.; Cai, Z.; Zhang, D.; Misra, D. Effect of deformation temperature on mechanical properties and deformation mechanisms of cold-rolled low C high Mn TRIP/TWIP steel. *Metals* **2018**, *8*, 476. [CrossRef]
5. Grzegorczyk, B.; Kozłowska, A.; Morawiec, M.; Muszyński, R.; Grajcar, A. Effect of deformation temperature on the Portevin-Le Chatelier effect in medium-Mn steel. *Metals* **2019**, *9*, 2. [CrossRef]
6. Kaar, S.; Krizan, D.; Schneider, R.; Béal, C.; Sommitsch, C. Effect of manganese on the structure-properties relationship of cold rolled AHSS treated by a quenching and partitioning process. *Metals* **2019**, *9*, 1122. [CrossRef]
7. Yang, Y.; Mi, Z.; Li, H.; Li, J.; Jiang, H. The impact of strain heterogeneity and transformation of metastable austenite on springback behavior in quenching and partitioning steel. *Metals* **2018**, *8*, 432. [CrossRef]
8. Sugimoto, K.; Sato, S.; Kobayashi, J.; Srivastava, A.K. Effects of Cr and Mo on mechanical properties of hot-forged medium carbon TRIP-aided bainitic ferrite steels. *Metals* **2019**, *9*, 1066. [CrossRef]
9. Sugimoto, K.; Hojo, T.; Srivastava, A.K. An overview of fatigue strength of case-hardening TRIP-aided martensitic steels. *Metals* **2018**, *8*, 355. [CrossRef]
10. Hojo, T.; Kobayashi, J.; Sugimoto, K.; Nagasaka, A.; Akiyama, E. Effects of alloying elements addition on delayed fracture properties of ultrahigh-strength TRIP-aided martensitic steels. *Metals* **2020**, *10*, 6. [CrossRef]

Article

Influence of the Quenching and Partitioning Process on the Transformation Kinetics and Hardness in a Lean Medium Manganese TRIP Steel

Simone Kaar [1,*], Reinhold Schneider [1], Daniel Krizan [2], Coline Béal [3] and Christof Sommitsch [3]

[1] Research and Development, University of Applied Sciences Upper Austria, Wels 4600, Austria; reinhold.schneider@fh-wels.at

[2] Research and Development Department, Business Unit Coil. Voestalpine Steel Division GmbH, Linz 4020, Austria; daniel.krizan@voestalpine.com

[3] Institute of Materials Science, Joining and Forming, Graz University of Technology, Graz 8010, Austria; coline.beal@tugraz.at (C.B.); christof.sommitsch@tugraz.at (C.S.)

* Correspondence: simone.kaar@fh-wels.at; Tel.: +43-50304-15-6250

Received: 27 February 2019; Accepted: 16 March 2019; Published: 19 March 2019

Abstract: The quenching and partitioning (Q&P) process of lean medium Mn steels is a novel approach for producing ultra-high strength and good formable steels. First, the steel is fully austenitized, followed by quenching to a specific quenching temperature (T_Q) in order to adjust an appropriate amount of initial martensite ($\alpha'_{initial}$). Subsequently, the steel is reheated to a partitioning temperature (T_P) in order to ensure C-partitioning from $\alpha'_{initial}$ to remaining austenite (γ_{remain}) and thus retained austenite (RA) stabilization. After isothermal holding, the steel is quenched to room temperature (RT), in order to achieve a martensitic-austenitic microstructure, where the meta-stable RA undergoes the strain-induced martensitic transformation by the so-called transformation induced plasticity (TRIP) effect. This paper systematically investigates the influence of the Q&P process on the isothermal bainitic transformation (IBT) kinetics in a 0.2C-4.5Mn-1.3Al lean medium Mn steel by means of dilatometry. Therefore, the Q&P annealing approach was precisely compared to the TRIP-aided bainitic ferrite (TBF) process, where the samples were directly quenched to the temperature of the IBT after full austenitization. The results indicated an accelerated IBT for the Q&P samples, caused by the formation of $\alpha'_{initial}$ during quenching below the martensite start (M_S) temperature. Furthermore, a significant influence of the annealing parameters, such as T_Q and T_P, was observed with regard to the transformation behavior. For further characterization, light optical microscopy (LOM) and scanning electron microscopy (SEM) were applied, showing a microstructure consisting of a martensitic-bainitic matrix with finely distributed RA islands. Saturation magnetization method (SMM) was used to determine the amount of RA, which was primarily depending on T_Q. Furthermore, the hardness according to Vickers revealed a remarkable impact of the annealing parameters, such as T_Q and T_P, on the predicted mechanical properties.

Keywords: Q&P; TRIP; lean medium Mn steel; transformation kinetics

---

1. Introduction

Both environmental and safety regulations force the automotive industry to the application of new steel grades [1,2]. In order to meet the challenging requirements concerning crashworthiness, formability, and reduction of CO_2 emissions, "Advanced High Strength Steels" (AHSS) are currently under development [3–5]. By the increase of the material strength without a deterioration of ductility, the sheet thickness can be downgauged, resulting in a significant weight reduction of the body structures [1,6]. Furthermore, the combination of both high strength and ductility is of vital importance in order to manufacture complex automotive parts. Therefore, a high research effort has been put into the further development of AHSS.

The 1st generation AHSS, being already in industrial application, is represented by Dual Phase (DP), TRIP, and Complex Phase (CP) steels [6]. These steels mainly have a multiphase structure, resulting in tensile strengths up to 1200 MPa and total elongations up to 40% [7–9]. Typically, they are characterized by $R_m \times A_{80}$ up to 20,000 MPa% [6]. The 2nd generation AHSS includes Twinning Induced Plasticity (TWIP), Nano-TWIP, Duplex, and Triplex steels, which are characterized by highly promising mechanical properties, especially an excellent combination of strength and ductility with $R_m \times A_{80}$ between 40,000 and 60,000 MPa% [10]. However, due to the high alloying costs and challenging processing they are hardly used in industrial applications [11].

Currently, recent research activities focus on the development of steels belonging to the 3rd generation AHSS, including the concepts of Q&P and medium Mn steels, to fill the property gap between the 1st and 2nd generation AHSS [9,12–14].

The Q&P process has been proposed as a promising approach for steel grades having the microstructure consisting of a carbon-depleted martensitic matrix and retained austenitic islands [15]. In this process, the steel is fully austenitized followed by quenching to a specific temperature below the M_S temperature, where the optimal amount of RA can be adjusted. Subsequent reheating and holding in the over-ageing region triggers the α' tempering, whereby the cementite formation during this stage will be significantly postponed by the Si and/or Al additions. Therefore, C can partition to remaining γ resulting in its appropriate stabilization upon final cooling to room temperature (RT) [16,17].

The medium Mn steels typically contain between 3 and 10 wt.% Mn and their microstructure consists of an ultrafine-grained ferritic matrix (with a typical grain size below 1 µm) and approximately 30 vol.% of RA. These steel grades are characterized by an excellent combination of tensile strength and total elongation achieving the product of $R_m \times A_{80}$ exceeding 30,000 MPa% [12,14,18].

In the present case, the concept of lean medium Mn Q&P steels combines both the approach of medium Mn steels and the Q&P process, leading to a microstructure consisting of a tempered martensitic matrix with an increased amount of RA islands. The RA can be stabilized by the combination of the C and Mn enrichment, ensuring an optimum strain-induced γ to α' transformation (TRIP-effect). Moreover, the presence of the hard C-depleted martensitic matrix ensures the superior performance of these steels in the forming operations such as sheet cutting, bending and hole expansion, since the ability to resist high local strains is linked to the hardness gradients in the microstructure [19].

In order to predict the amount of RA depending on T_Q, the constrained carbon equilibrium (CCE) model proposed by Speer et al. [20] can be applied. This simplified model allows the calculation of the C-content in γ and therefore the prediction of the amount of RA under three main conditions: (1) All of the C partitions to γ and the partitioning kinetics are ignored, meaning the partitioning is already complete; (2) no phase boundary movement during the partitioning process; and (3) no competing reactions like carbide or α_B formation take place during the Q&P process [20].

However, the decomposition from γ to α_B during the C-partitioning step has been observed in some instances [21–23]. Clarke et al. [24] showed that the formation of carbide-free bainite (α_B) during the partitioning process led to a reduced RA fraction in the Q&P samples compared to the amounts predicted by the CCE-model. However, in some cases, the formation of α_B can contribute to the stabilization of γ by its C-enrichment.

Therefore, in the present paper, the transformation behavior of a 0.2-4.5Mn-1.3Al steel was studied for both Q&P and TBF processes in order to evaluate the influence of the presence of $\alpha'_{initial}$ on IBT. Several annealing parameters were varied (T_Q, T_P), in order to examine their influence on the transformation kinetics and thus the volume fraction of the individual constituents, i.e., α', α_B, and RA, in the final microstructure.

2. Materials and Methods

Table 1 shows the chemical composition of the investigated steel grade in wt.%. The steel was melted in a medium frequency induction furnace and cast under laboratory conditions in an ingot of 80 kg. First the material was hot rolled to a thickness of 4 mm, followed by tempering in a batch-annealing-like furnace at a temperature of 550 °C for 16 h. Finally, the material was cold rolled to a thickness of 1 mm.

Table 1. Chemical composition of the investigated steel Fe-C-Mn-Al in wt.%.

C	Mn	Al	Si
0.20	4.49	1.30	0.04

The transformation behavior and the influence of the different annealing parameters were investigated by means of dilatometry using a Bähr 805 A/D dilatometer. Therefore, specimens with dimensions of $10 \times 4 \times 1$ mm^3 were produced by wire-electrical discharge machining. Figure 1 shows the applied time-temperature schedules for the Q&P (a) and the TBF (b) heat-treatments, adapted to suit an industrially feasible continuous annealing line. All samples were fully austenitized at a T_{an} of 900 °C for 120 s (t_{an}) with a heating rate HR$_1$ of 10 K/s. For the Q&P process, the austenitization was followed by quenching to various T_Q in the range of 130 °C–330 °C with a 20 °C step using a cooling rate CR$_1$ of 50 K/s. The T_Q was held for 10s and subsequently the samples were reheated with 20 K/s (HR$_2$) to a certain T_P, which was varied between 350, 400, and 450 °C and held for 600 s (t_p), respectively. In contrast, the TBF samples were directly cooled to a T_B of 350, 400, or 450 °C and held for 600 s (t_B). Finally, both Q&P and TBF samples were cooled to RT with a cooling rate CR$_2$ of 50 K/s.

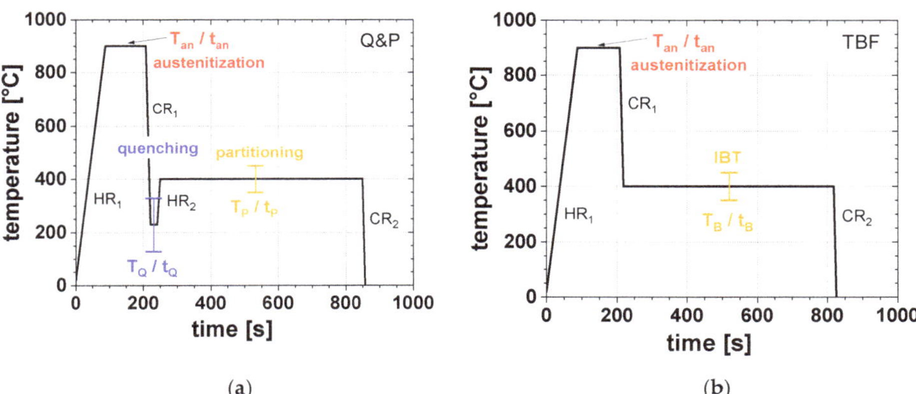

(a)   (b)

Figure 1. Time-temperature regimes used for (**a**) the quenching and partitioning (Q&P) and (**b**) the transformation induced plasticity (TRIP) aided bainitic ferrite (TBF) heat-treatments.

The microstructure was characterized by means of light optical microscopy (LOM) using LePera etching [25]. Furthermore, the samples were electrochemically polished for the SEM investigations. Vickers hardness testing (HV1) was performed on polished samples using an Emco Test DuraScan 20 device. The volume fraction of RA was determined by the use of SMM [26]. Furthermore, the RA content depending on the T_Q was calculated using the CCE-model proposed by Speer et al. [20]. Therefore, first the volume fraction of α' formed during quenching to T_Q was determined by the Koistinen-Marburger (KM) equation, which can be used to predict the γ to α' transformation rate and is given by [27]:

$$f_M = 1 - e^{-0.011(M_S - T_Q)}$$

Here, f_M is the fraction of α' transformed during quenching from the γ-region to the T_Q. M_S is the martensite start temperature, which was calculated according to Mahieu et al. [28]:

$$M_S = 539 - 423\text{C} - 30.4\text{Mn} - 7.5\text{Si} + 30\text{Al}$$

Figure 2a displays the phase fractions of α' and γ determined using the CCE-model and Figure 2b represents the related Q&P heat-treatment. Both, KM- and M_S-formula are applied twice in the CCE-model. First, the M_S temperature of the initial γ is calculated, followed by the determination of the α' fraction formed during quenching to a certain T_Q using the KM equation (blue line). The orange curve represents the remaining initial γ after quenching. Subsequently, the M_S temperature of the C-enriched γ after the partitioning step is calculated assuming full C-partitioning from α'. Given that, using the KM equation once again, the amount of α' formed upon final cooling to RT is calculated (red line). The resulting RA fraction as a function of the T_Q is indicated in green. It is evident that the largest fraction of γ can be retained at that T_Q where no fresh α' is formed upon final cooling. However, it must be taken into account that this applied CCE-model does not consider a phase transformation from γ to α_B during isothermal holding at T_P. Since this contribution focusses on the investigation of the bainitic transformation kinetics during Q&P processing, it can be expected that the amount of retained austenite stabilized to RT will be smaller, compared to the calculated one.

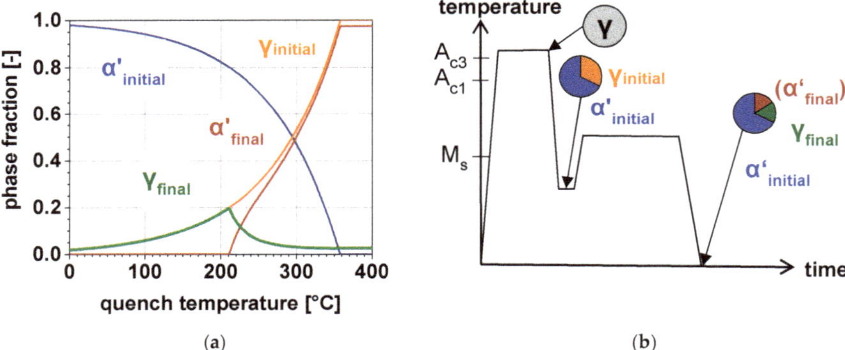

Figure 2. (**a**) Calculated phase fractions of martensite (α') and austenite (γ) depending on the T_Q and (**b**) the related Q&P heat-treatment with the schematic microstructural evolution; the phase fractions are indicated for the initial quench to T_Q and for the final quench to room temperature (RT).

3. Results

3.1. Transformation Behavior

The annealing parameters, especially T_Q and T_P, remarkably influenced the transformation behavior of the investigated steel. Figure 3 shows the influence of the T_P-variation in a range of 350–450 °C on the dilatometric curves using the Q&P heat-treatment with a T_Q of 310 °C (Figure 3a) compared to the TBF cycles (Figure 3b). After full austenitization at 900 °C, the dilatometric samples were cooled to T_Q or T_B, depending on the applied heat-treatment. In the case of the Q&P regime, $\alpha'_{initial}$ was formed during quenching to T_Q, since the M_S temperature was approximately 325 °C (Figure 3a). Although the CCE-model, describing the microstructural development during Q&P processing, assumes full C-partitioning from γ to α' without the occurrence of any phase transformations during isothermal holding at T_P, in the present case after reheating to T_P, γ partially transformed to α_B during isothermal holding, accompanied by a linear expansion visible in the dilatometric curves. Further, it is obvious that the amount of α_B described by the volume expansion during isothermal holding decreased with increasing T_P. As a result, T_P directly influenced the formation of α'_{final} during cooling to RT. For the Q&P sample heat-treated at a T_P of 350 °C, no α'_{final} was formed, whereas at an increased T_P of 400 and 450 °C, a $\gamma \rightarrow \alpha'_{\text{final}}$ transformation was observed by a deviation from the linearity of the dilatometric curves during final cooling. In contrast, for the TBF samples, the largest amount of α_B was formed at a T_B of 400 °C, whereas its smallest amount was contemplated at 450 °C. The correlation between the amount of α_B and α'_{final} was proven for the TBF regime as well. Thus, the largest amount of α'_{final} was transformed at a T_B of 450 °C, whereas the lowest α' formation was obtained at 400 °C.

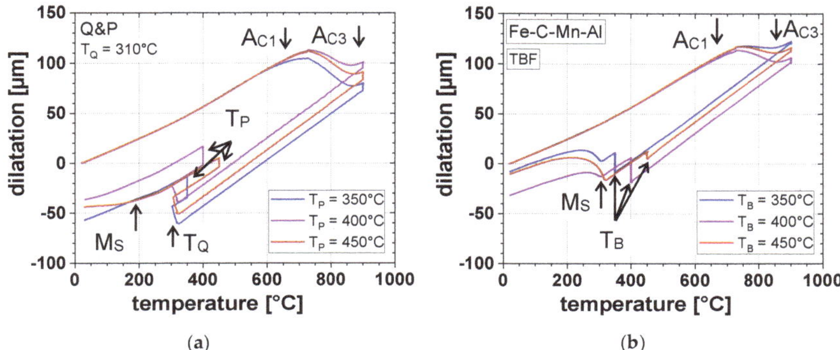

Figure 3. Dilatometric curves at different T_P (T_B) of (**a**) Q&P heat-treatment (T_Q = 310 °C) and (**b**) TBF heat-treatment.

The influence of T_P or T_B on the α_B formation during isothermal holding for the Q&P (T_Q = 310 °C) and TBF process is displayed in detail in Figure 4. For both the Q&P and TBF regimes, an increasing T_P and T_B led to accelerated IBT kinetics. However, with an increase of T_P and T_B, the α_B formation saturated already after shorter times. In the case of the Q&P samples, this behavior led to the largest amount of α_B at a T_P of 350 °C, whereas with increasing T_P, lower α_B fractions were formed during isothermal holding. On the one hand, the IBT in the TBF regime was remarkably slower compared to that of the Q&P process. On the other hand, the IBT did not saturate at a T_B of 350 °C and 400 °C, even after a t_B of 600 s.

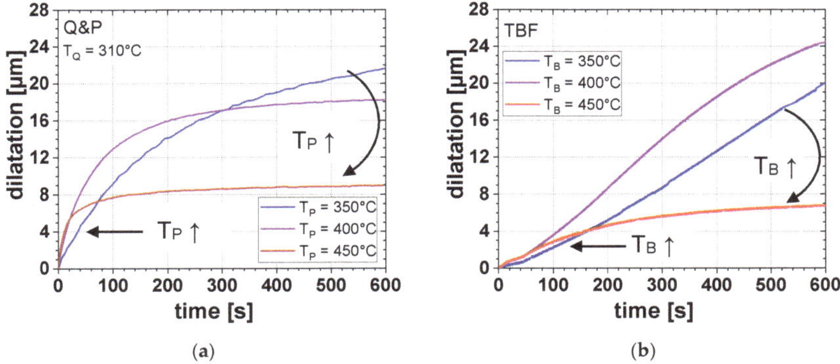

Figure 4. Dilatation due to α_B formation as a function of isothermal holding time at different T_P (T_B) for (**a**) Q&P steel (T_Q = 310 °C) and (**b**) TBF steel.

Furthermore, an influence of T_Q on the transformation behavior could be determined by means of dilatometry (Figure 5). With increasing T_Q, the amount of α'_{initial} decreased, whereas the amount of

α_B formed during isothermal holding at T_P increased. However, too high T_Q led to the formation of α'_{final} during final cooling to RT, especially at a T_P of 450 °C.

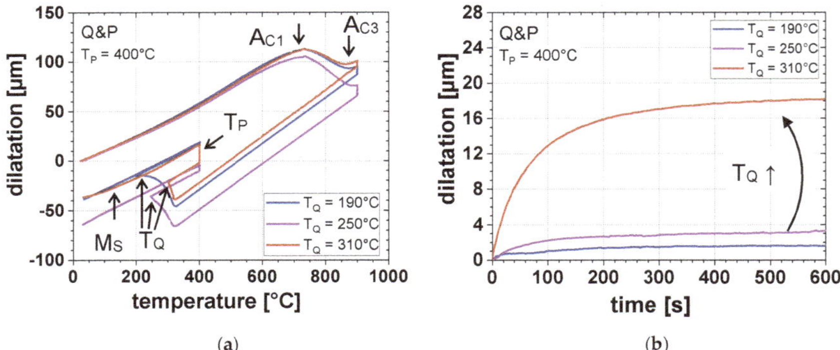

(a)   (b)

Figure 5. (a) Dilatometric curves at different T_Q (T_P = 400 °C) and (b) dilatation due to α_B formation as a function of isothermal holding time at different T_Q (T_P = 400 °C).

3.2. Microstructure

The annealing parameters, especially T_Q, significantly influenced the microstructural constitution of the investigated samples. Figure 6a–c shows the LOM images of the Q&P samples, quenched to a temperature of 190, 250, and 310 °C, respectively. In comparison, the microstructure of the sample annealed in the TBF regime is shown in Figure 6d. All four conditions were heat-treated at a T_P (T_B) of 400 °C. For the Q&P steels, the matrix consisted of a mixture of $\alpha'_{initial}$ and α_B, which appeared as bluish and brownish areas in the micrographs. Furthermore, RA and/or α'_{final} islands (white and brownish areas) could be observed as finely distributed in the matrix. At the lowest T_Q of 190 °C, the secondary phase only consisted of RA, whereas the increase of T_Q led to the pronounced formation of α'_{final}. The microstructure, especially at a T_Q of 250 and 310 °C exhibited a banded structure, resulting from the segregation of alloying elements such as Mn during casting and subsequent solidification. In contrast, the microstructure of the TBF samples consisted of a ferritic-bainitc matrix with small amounts of finely distributed RA islands, surrounded by large fractions of α'_{final}.

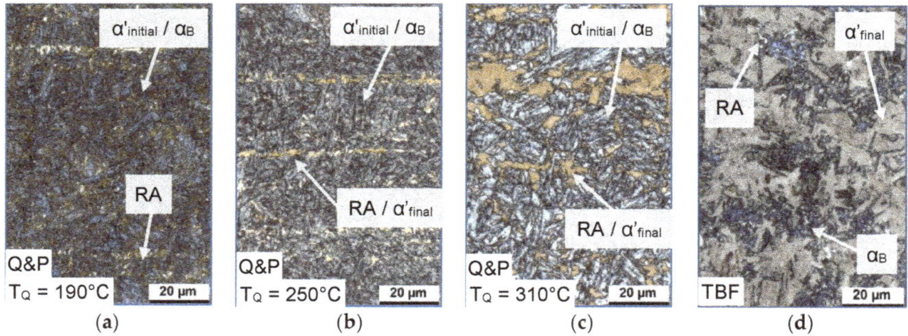

(a)   (b)   (c)   (d)

Figure 6. Light optical microscopy (LOM) images of the (**a**–**c**) Q&P samples (T_Q = 190, 250, and 310 °C) and (**d**) TBF sample at a T_P (T_B) of 400 °C (magnification 1000×).

For higher resolution investigations, the microstructure was further characterized by SEM. The SEM images of the Q&P samples quenched to a T_Q of 190, 250, and 310 °C are displayed in Figure 7a–c, whereas in Figure 7d the microstructure after the TBF heat-treatment is shown for comparison. In the case of the Q&P samples, the microstructure contained cementite precipitates with an obvious triaxial alignment, which means that the precipitations are arranged in an angle of 120°. This indicates the presence of tempered α'_{initial}. Furthermore, the fraction of α_B in the matrix rose along with the amount of α'_{final} with increasing T_Q. In comparison, the microstructure of the TBF steel consisted of a α_B-matrix with large amounts of α'_{final} and low fractions of RA.

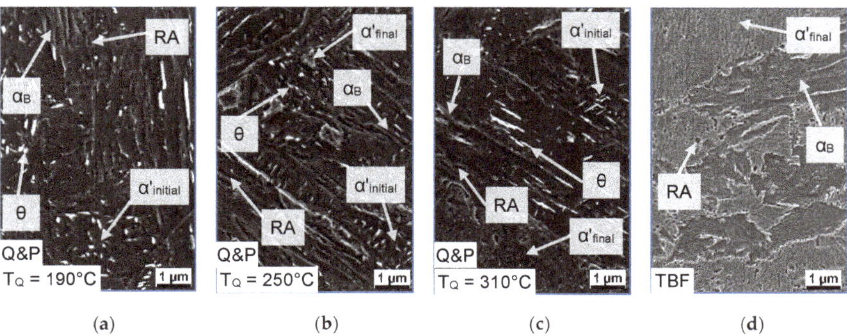

(**a**)   (**b**)   (**c**)   (**d**)

Figure 7. SEM images of the (**a**–**c**) Q&P samples (T_Q = 190, 250, and 310 °C) and (**d**) TBF sample at a T_P (T_B) of 400 °C (magnification 5000×).

Figure 8 depicts the amount of RA, determined by SMM, for the TBF samples (dotted lines) and the Q&P samples at a T_P of 350, 400, and 450 °C, respectively. In addition, the RA fraction calculated according to the CCE-model proposed by Speer et al. [20] is presented in the diagram. The CCE-model proposed a RA maximum (RA$_{\text{max}}$) of 19.7 vol.% at a T_Q of 210 °C. In accordance with the model calculations, the experiments also confirmed the shape of the RA evolution as a function of T_Q. At lower T_Q the formation of high amounts of α'_{initial} led to lower RA fractions. By a further increase of the T_Q, the RA$_{\text{max}}$ was achieved followed by a decrease of the RA. This was due to its lower stabilization and therefore the formation of α'_{final} upon cooling to RT. Furthermore, the experiments showed an influence of T_P on the amount of RA: with increasing T_P, RA$_{\text{max}}$ rose from 14.8 to 19.0 vol.%. In this context, at the T_P of 450 °C, the amount of RA correlated well with the model calculations. On the contrary, at the T_P of 350 °C and 400 °C, a lower amount of RA was achieved compared to the model predictions. Moreover, the T_Q at which the RA$_{\text{max}}$ occurred, was shifted to higher temperatures by up to 80 °C compared to the CCE-model. It is evident that the RA contents of the TBF samples were lower compared to the Q&P samples: 7.1 vol.% RA at a T_B of 350 °C, 9.7 vol.% RA at T_B = 400 °C, and 3.6 vol.% RA at T_B = 450 °C.

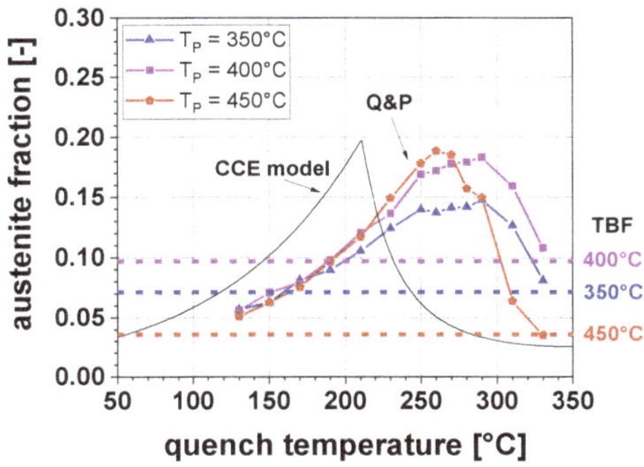

Figure 8. Retained austenite (RA) content as a function of T_Q for a T_P (T_B) of 350, 400, and 450 °C.

Figure 9 represents microstructure charts, where the phase fractions are plotted as a function of T_Q for different T_P. In addition, the bar charts, situated close to the right rim of each diagram, depict the microstructural constitution for the TBF steel. In case of the Q&P samples, the decrease of T_Q led to a considerable increase of $\alpha'_{initial}$. As a consequence, a lower amount of α_B was formed during IBT, regardless of T_P. However, particularly at higher T_Q, a vivid influence of T_P is obvious: the increase of T_P led to a decreasing fraction of α_B, resulting in the formation of α'_{final} upon cooling to RT. Therefore, the RA content steadily rose with increasing T_Q until the onset of the formation of α'_{final} occured. For the TBF samples, it is evident that the largest amount of α_B was formed at a T_B of 400 °C as already shown by dilatometry (Figure 4b). Due to the lowest α_B fraction at a T_B of 450 °C, the largest amount of γ transformed to α'_{final} during final cooling, resulting in the lowest amount of RA, followed by the samples heat-treated at a T_B of 350 °C and 400 °C.

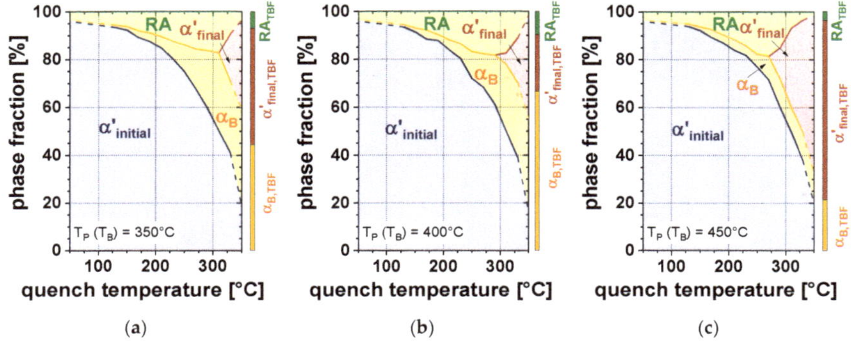

Figure 9. Phase fraction as a function of T_Q for a T_P (T_B) of (**a**) 350 °C, (**b**) 400 °C, and (**c**) 450 °C.

3.3. Hardness

The results of the hardness measurements according to Vickers are plotted as a function of T_Q in Figure 10. When the T_Q was in the range of 130 to 250 °C, both increasing T_Q and T_P led to a slightly decreasing hardness. However, at T_Q exceeding 250 °C, the hardness rose with T_Q, especially at a T_P of 450 °C. For this reason, in case of the Q&P samples, the hardness was between approximately 350 and 460 HV1. In contrast, for the TBF samples the lowest hardness of 445 HV1 was measured at a T_B of 400 °C, whereas at a T_B of 350 and 450 °C the hardness reached approximately 470 and 475 HV1, respectively.

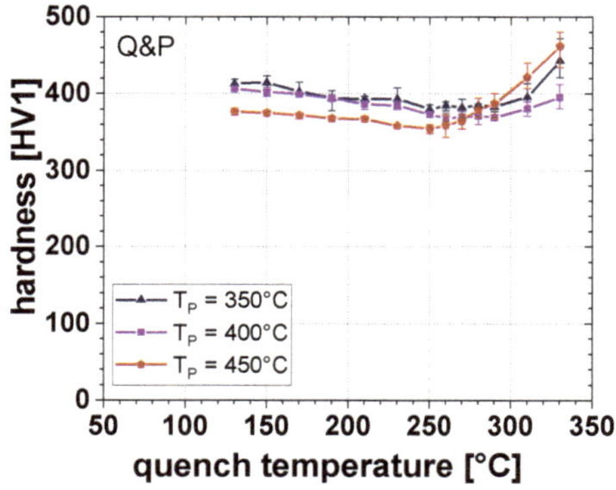

Figure 10. Vickers hardness as a function of T_Q for a T_P of 350, 400, and 450 °C.

4. Discussion

The time-temperature-transformation (TTT) diagrams for isothermal holding at different T_P (T_B) depict the comparison between the Q&P heat-treatment (Figure 11a–c) at different T_Q (190, 250, and 310 °C) and the TBF processing route (Figure 11d). It is obvious that the application of the Q&P process resulted in an evident acceleration of the α_B formation. This is in agreement with Wang et al. [29], since the presence of $\alpha'_{initial}$ pronouncedly accelerated the IBT due to the enhanced nucleation rate by the presence of geometrically necessary dislocations. At higher T_Q, the amount of α_B significantly increased, in particular at the lower T_P of 350 and 400 °C. According to Smanio and Sourmail [30], this relation can be explained by the fact that at higher T_Q lower amounts of $\alpha'_{initial}$ were formed, which was also confirmed by the LOM and SEM investigations (Figures 6 and 7). As a result, a larger amount of γ was present in the microstructure at the onset of isothermal holding, from which a larger amount of α_B could be formed, compared to lower T_Q. For this reason, the largest α_B formation was observed in the TBF samples, where especially a lower T_B of 350 °C led to increased α_B fractions up to 80 vol.% after the investigated holding time. Furthermore, a general impact of the T_P (T_B) on the transformation kinetics was found: especially at a T_P (T_B) of 400 and 450 °C the transformation kinetics was evidently accelerated due to the faster C diffusion into the remaining γ. In general, the increase of T_P and T_B led to lower fractions of α_B, which is in analogy with the T_0-concept [31]. Thereby, the difference between the Gibbs free energies of γ and α decreases with an increase of temperature. As a consequence, the driving force for α_B formation diminishes with temperature.

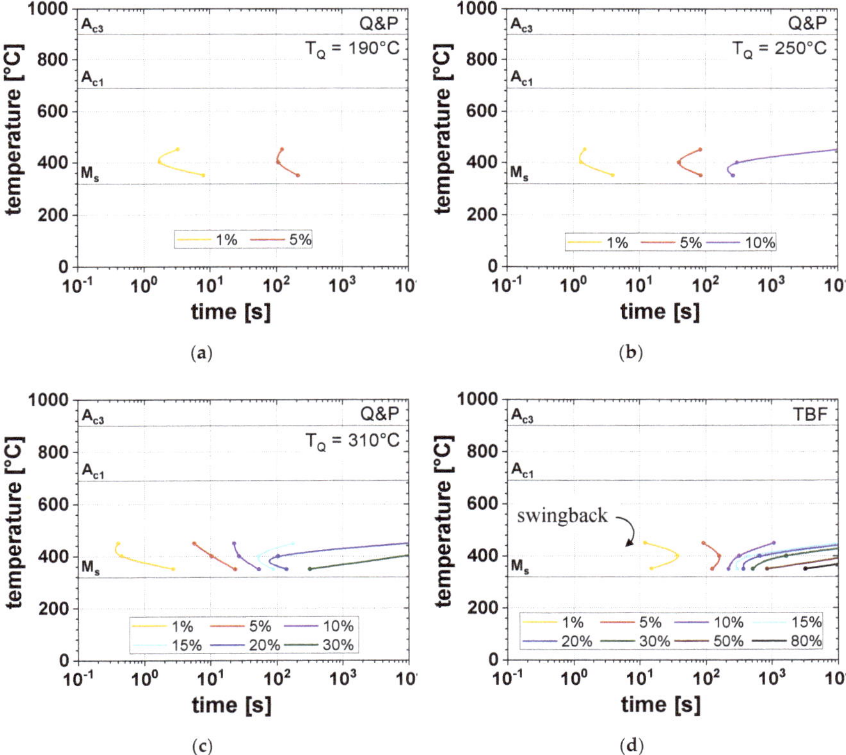

Figure 11. Time-temperature-transformation (TTT) diagrams for (**a–c**) Q&P steels (T_Q = 190, 250, and 310 °C) and (**d**) TBF steels.

In the case of the TBF samples, the swing back phenomenon could be contemplated, which is in accordance with Oka and Okamoto [32]. This effect describes the accelerated formation of α_B due to the formation of thin-plate isothermal α' in a temperature range just above M_S. Therefore, at a T_B of 350 °C, the onset of the α_B formation was observed at shorter times compared to a T_B of 400 °C.

Considering the RA investigations, a significant difference between the calculated CCE-model and experimental data was determined. This was particularly apparent for the amount of RA and optimal T_Q. In analogy to the present results, these observations have already been reported in the literature [24,33,34]. The CCE-model assumes full C-partitioning from $\alpha'_{initial}$ into the remaining γ. However, in the present study the microstructural investigations have shown the presence of cementite precipitates in the tempered $\alpha'_{initial}$ (Figure 7), which contradicts this assumption. The insufficient C-partitioning led to a reduced C-content in the remaining γ, resulting in a lower amount of RA due to the formation of α'_{final}, caused by the lower chemical RA stability. In particular, for the T_P of 350 °C, the RA_{max} was remarkably lower than the predicted one. According to Clarke et al. [24], the α_B formation, which predominantly occurred at a T_P of 350 °C in the present study, led to a pronounced decrease in RA as the aftermath of the reduction of untransformed γ during isothermal holding.

Furthermore, the microstructural changes were reflected in hardness according to Vickers, as well (Figures 9 and 10). With increasing RA, content hardness steadily decreased, resulting in the lowest hardness for those samples containing the largest amount of RA. Since with increasing T_P the volume fraction of RA stabilized to RT rose, the lowest hardness values could be obtained at $T_P = 450$ °C. However, the increased RA content at higher T_P led to the lower chemical RA stability, and thus to an intensified formation of α'_{final}. Therefore, especially at $T_P = 450$ °C, hardness drastically increased at T_Q, exceeding that temperature where the maximum RA content was measured. This is in agreement with [35], where a comparable correlation between the microstructural constituents (i.e., RA and α'_{final}) and resulting hardness was found for lean medium Mn steels.

The evident shift of the RA_{max} to higher T_Q than predicted could be explained by the fact that γ is stabilized by both chemical and mechanical factors, whereas the CCE-model only takes the chemical stabilization into account [36]. Furthermore, the CCE-model includes two empirical formulas, describing the α' kinetics and the M_S temperature. According to Kim et al. [37], slight differences in the chemical composition or the initial microstructure could therefore lead to deviations from the model, as well. In the present contribution, small deviations between the M_S temperature calculated according to Mahieu et al. ($M_S = 350$ °C) and the M_S temperature determined by means of dilatometry ($M_S = 325$ °C, Figure 5a) could be observed, as well. Therefore, an adjustment of the applied M_S-formula with an improved validity for increased Mn and C contents is intended for future investigations. Nevertheless, the CCE-model proposed by Speer et al. [20] is a vital tool for the first estimation of applicable annealing parameters in the case of Q&P steels.

5. Conclusions

The results of the present contribution give conclusive evidence that the Q&P process significantly influences the transformation behavior of the investigated lean medium Mn steel. Both T_Q and T_P must be carefully adjusted, since they have a substantial impact on the microstructural evolution during steel processing.

The main findings of the present study are as follows:

- The Q&P heat-treatment accelerated the transformation kinetics during isothermal holding in the partitioning step due to the presence of $\alpha'_{initial}$, which acted as nucleation sites for the α_B formation.
- For both Q&P and TBF grades, the increase of T_P (T_B) resulted in lower amounts of α_B, although the transformation kinetics was accelerated.
- With increasing T_Q, smaller amounts of $\alpha'_{initial}$ and larger amounts of α_B were formed, leading to an increased RA content. At too high T_Q, α'_{final} was formed, resulting in a sharp decrease in

RA. Therefore, the microstructure of the Q&P samples consisted of tempered $\alpha'_{initial}$, RA, and at higher T_Q partially of α_B and α'_{final}.

- The microstructure of the TBF samples with the present lean medium Mn composition consisted of a ferritic-bainitic matrix with large amounts of α'_{final} and low fractions of RA.
- The comparison of the CCE-model and the experimental data showed a significant deviation of the maximum RA content, in particular at lower T_P. Furthermore, the optimum T_Q was shifted to higher temperatures compared to the CCE-model predictions.

Author Contributions: Conceptualization, R.S., D.K., and C.S.; methodology, S.K.; software, S.K.; validation, D.K. and R.S.; formal analysis, S.K.; investigation, S.K.; resources, D.K., R.S., and S.K.; data curation, S.K., R.S., D.K., C.B., and C.S.; writing—original draft preparation, S.K.; writing—review and editing, R.S., D.K., C.B., and C.S.; visualization, S.K.; supervision, R.S., D.K., C.B., and C.S.; project administration, D.K., R.S., and C.S.; funding acquisition, D.K.

Funding: This research was funded by the Austrian Research Promotion Agency (FFG), grant number 860188, "Upscaling of medium Mn-TRIP steels".

Conflicts of Interest: The authors declare no conflict of interest. The funders had no role in the design of the study; in the collection, analyses, or interpretation of data; in the writing of the manuscript, or in the decision to publish the results.

References

1. Kwon, O.; Lee, K.; Kim, G.; Chin, K. New trends in advanced high strength steel—Developments for automotive application. *Mater. Sci. Forum* **2010**, *638–642*, 136–141. [CrossRef]
2. Spenger, F.; Hebesberger, T.; Pichler, A.; Krempaszky, C.; Werner, E.; Doppler, C. AHSS steel grades: Strain hardening and damage as material design criteria. In Proceedings of the International Conference on New Developments in Advanced High Strength Sheet Steels, Orlando, FL, USA, 15–18 June 2008; pp. 39–49.
3. Jacques, P.J.; Petein, A.; Harlet, P. Improvement of mechanical properties through concurrent deformation and transformation: New steels for the 21st century. In Proceedings of the International Conference on TRIP-aided High Strength Ferrous Alloys, Aachen, Germany, 19–21 June 2002; pp. 281–285.
4. Hashimoto, K.; Yamasaki, M.; Fujimura, K.; Matsui, T.; Izumiya, K. Global CO_2 recycling—Novel materials and prospect for prevention of global warming and abundant energy supply. *Mater. Sci. Eng. A* **1999**, *267*, 200–206. [CrossRef]
5. Zaefferer, S.; Ohlert, J.; Bleck, W. A study of microstructure, transformation mechanisms and correlation between microstructure and mechanical properties of a low alloyed TRIP steel. *Acta Mater.* **2004**, *52*, 2765–2778. [CrossRef]
6. Matlock, D.; Speer, J.; De Moor, E.; Gibbs, P. Recent developments in advanced high strength steels for automotive applications: An overview. *JESTECH* **2012**, *15*, 1–12.
7. De Cooman, B.C. Structure-properties relationship in TRIP steels containing carbide-free bainite. *Solid State Mater. Sci.* **2004**, *8*, 285–303. [CrossRef]
8. Hairer, F.; Krempaszky, C.; Tsipouridis, P.; Werner, E.; Satzinger, K.; Hebesberger, T.; Pichler, A. Effects of heat treatment on microstructure and mechanical properties of bainitic single- and complex-phase steel. *Proc MS&T* **2009**, *2009*, 1391–1401.
9. Samek, L.; Krizan, D. Steel—Material of choice for automotive lightweight applications. In Proceedings of the International Conference on Metals, Brno, Czech Republic, 23–25 May 2012; pp. 6–12.
10. De Cooman, B.C.; Chin, K.; Kim, J. High Mn TWIP steels for automotive applications. In *New Trends and Developments in Automotive System Engineering*; Chiaberge, M., Ed.; InTech: Rijeka, Croatia, 2010; pp. 101–128.
11. Bracke, L.; Verbeken, K.; Kestens, L.; Penning, J. Microstructure and texture evolution during cold rolling and annealing of a high Mn TWIP steel. *Acta Mater.* **2009**, *57*, 1512–1524. [CrossRef]
12. Steineder, K.; Schneider, R.; Krizan, D.; Bèal, C.; Sommitsch, C. Microstructural evolution of two low-carbon steels with a medium manganese content. In Proceedings of the 2nd HMnS Conference, Aachen, Germany, 31 August–4 September 2014; pp. 351–354.
13. De Cooman, B.C.; Speer, J. Quench and partitioning steel: A new AHSS concept for automotive anti-intrusion applications. *Steel Res. Int.* **2006**, *77*, 634–640. [CrossRef]

14. Steineder, K.; Krizan, D.; Schneider, R.; Bèal, C.; Sommitsch, C. On the microstructural characteristics influencing the yielding behavior of ultra-fine grained medium-Mn steels. *Acta Mater.* **2017**, *139*, 39–50. [CrossRef]

15. Speer, J.; Matlock, D.; De Cooman, B.C.; Schroth, J. Carbon partitioning into austenite after martensite transformation. *Acta Mater.* **2003**, *51*, 2611–2622. [CrossRef]

16. Speer, J.; Aussunção, F.; Matlock, D.; Edmonds, D. The "quenching and partitioning" process: Background and recent progress. *Mater. Res.* **2005**, *8*, 417–423. [CrossRef]

17. Edmonds, D.; He, K.; Rizzo, F.; De Cooman, B.C.; Matlock, D.; Speer, J. Quenching and partitioning martensite—A novel steel heat treatment. *Mater. Sci. Eng. A* **2006**, *438–440*, 25–34. [CrossRef]

18. Arlazarov, A.; Gouné, M.; Bouaziz, O.; Hazotte, A.; Petitgand, G.; Barges, P. Evolution of microstructure and mechanical properties of medium Mn steels during double annealing. *Mater. Sci. Eng. A* **2012**, *542*, 31–39. [CrossRef]

19. Huyghe, P.; Dépinoy, S.; Caruso, M.; Mercier, D.; Georges, C.; Malet, L.; Godet, S. On the effect of Q&P processing on the stretch-flange-formability of 0.2C Ultra-high Strength Steel Sheets. *ISIJ Int.* **2018**, *58*, 1341–1350.

20. Speer, J.; Streicher, A.; Matlock, D.; Rizzo, F. Quenching and partitioning: A fundamentally new process to create high strength TRIP sheet microstructures. In Proceedings of the Austenite Formation and Decomposition, Chicago, IL, USA, 9–12 November 2003; pp. 505–522.

21. Van Bohemen, S.; Santofimia, M.; Sietsma, J. Experimental evidence for bainite formation below M_S in Fe-0.66C. *Scr. Mater.* **2009**, *58*, 488–491. [CrossRef]

22. Somani, M.; Porter, D.; Karjalainen, L.; Misra, R. On various aspects of decomposition of austenite in a high-silicon steel during quenching and partitioning. *Metall. Mater. Trans. A* **2013**, *45*, 1247–1257. [CrossRef]

23. Kim, D.; Speer, J.; De Cooman, B.C. Isothermal transformation of a CMnSi steel below the M_S temperature. *Metall. Mater. Trans. A* **2011**, *42*, 1575–1585. [CrossRef]

24. Clarke, A.; Speer, J.; Miller, M.; Hackenberg, R.; Edmonds, D.; Matlock, D.; Rizzo, F.; Clarke, K.; De Moor, E. Carbon partitioning to austenite from martensite or bainite during the quench and partition (Q&P) process: A crictial assessment. *Acta Mater.* **2008**, *56*, 16–22.

25. Zakerinia, H.; Kermanpur, A.; Najafizadeh, A. Color metallography: A suitable method for characterization of martensite and bainite in multiphase steels. *Int. J. ISSI* **2009**, *6*, 14–18.

26. Wirthl, E.; Pichler, A.; Angerer, R.; Stiaszny, P.; Hauzenberger, K.; Titovets, Y.F.; Hackl, M. Determination of the volume amount of retained austenite and ferrite in small specimens by magnetic measurements. In Proceedings of the International Conference on TRIP-aided high strength ferrous alloys, Ghent, Belgium, 19–21 June 2002; pp. 61–64.

27. Koistinen, D.; Marburger, R. A general equation prescribing the extent of the austenite-martensite transformation in pure iron-carbon alloys and plain carbon steels. *Acta Metall.* **1959**, *7*, 59–60. [CrossRef]

28. Mahieu, J.; Maki, J.; De Cooman, B.C.; Claessens, S. Phase transformation and mechanical properties of Si-free CMnAl transformation-induced plasticity-aided steel. *Metall. Mater. Trans. A* **2002**, *33*, 2573–2580. [CrossRef]

29. Wang, G.; Chen, S.; Liu, C.; Wang, C.; Zhao, X.; Xu, W. Correlation of isothermal bainite transformation and austenite stability in quenching and partitioning steels. *J. Iron Steel Res. Int.* **2017**, *24*, 1095–1103.

30. Smanio, V.; Sourmail, T. Effect of partial martensite transformation on bainite reaction kinetics in different 1%C steels. *Solid State Phenom.* **2011**, *172–174*, 821–826. [CrossRef]

31. Bhadeshia, H.K.D.H.; Honeycombe, R. The bainite reaction. In *Steels—Microstructure and Properties*, 3rd ed.; Edward Arnold: London, UK, 2006; pp. 121–124.

32. Oka, M.; Okamoto, H. Swing back in kinetics near M_S in hypereutectoid steels. *Metall. Trans. A* **1988**, *19*, 447. [CrossRef]

33. Steineder, K.; Schneider, R.; Krizan, D.; Bèal, C.; Sommitsch, C. Investigation on the microstructural evolution in a medium-Mn steel (X10Mn5) after intercritical annealing. *HTM* **2015**, *70*, 19–25. [CrossRef]

34. De Moor, E.; Lacroix, S.; Clarke, A.; Penning, J.; Speer, J. Effect of retained austenite stabilized via quench and partitioning on the strain hardening of martensitic steels. *Metall. Mater. Trans. A* **2008**, *39*, 2586–2595. [CrossRef]

35. Grajcar, A.; Skrzypczyk, P.; Kozlowska, A. Effects of temperature and time of isothermal holding on retained austenite stability in medium-Mn steels. *Appl. Sci.* **2018**, *8*, 2156. [CrossRef]

36. Jimenez-Melero, E.; Van Dijk, N.; Zhao, L.; Sietsma, J.; Offerman, S.; Wright, J.; Van der Zwaag, S. Martensitic transformation of individual grains in low-alloyed TRIP steels. *Scr. Mater.* **2007**, *56*, 421–424. [CrossRef]
37. Kim, S.; Lee, J.; Barlat, F.; Lee, M.G. Transformation kinetics and density models of quenching and partitioning (Q&P) steels. *Acta Mater.* **2016**, *109*, 394–404.

Article

Quantitative Analysis of Microstructure Evolution in Hot-Rolled Multiphase Steel Subjected to Interrupted Tensile Test

Adam Grajcar [1],*, Aleksandra Kozłowska [1], Krzysztof Radwański [2] and Adam Skowronek [1]

[1] Department of Engineering Materials and Biomaterials, Faculty of Mechanical Engineering, Silesian University of Technology, 18A Konarskiego Street, 44-100 Gliwice, Poland; aleksandra.kozlowska@polsl.pl (A.K.); adam.skowronek@polsl.pl (A.S.)

[2] Łukasiewicz Research Network-Institute for Ferrous Metallurgy, 12-14 K. Miarki Street, 44-100 Gliwice, Poland; kradwanski@imz.pl

* Correspondence: adam.grajcar@polsl.pl; Tel.: +48-322-372-933

Received: 4 November 2019; Accepted: 30 November 2019; Published: 3 December 2019

Abstract: A quantitative analysis of the microstructure evolution in thermomechanically processed Si-Al multiphase steel with Nb and Ti microadditions was performed in the study. The tendency of strain-induced martensitic transformation of retained austenite was analyzed during a tensile test interrupted at incremental strain levels. Optical micrographs and electron backscatter diffraction (EBSD) maps were obtained at each deformation step. The quantitative analysis of the martensitic transformation progress as a function of strain was performed. The results showed that the stability of retained austenite is mostly related to its grain size and morphology. Large, blocky-type grains of retained austenite located in a ferritic matrix easily transformed into martensite during an initial step of straining. The highest mechanical stability showed small austenitic grains and thin layers located in bainitic islands. It was found that the extent of martensitic transformation decreased as the deformation level increased.

Keywords: high-strength steel; retained austenite; TRIP effect; strain-induced martensitic transformation; multiphase microstructure; microalloying

---

1. Introduction

A microstructure of transformation-induced plasticity (TRIP)-assisted steels typically contain ferrite, bainite, retained austenite, and sometimes a small fraction of martensite. Retained austenite is a key microstructural constituent due to its strain-induced transformation to martensite. This enables obtaining beneficial combinations of strength and ductility. Steels sheets in the automotive industry are subjected to multi-step forming operations, in which some fraction of retained austenite is transformed into the martensite. Thus, the total amount of retained austenite available for the transformation reduces as deformation level increases [1,2]. Therefore, it is important to characterize the microstructure evolution of TRIP steels during incremental deformation. This enables predicting optimal mechanical properties of final products and monitoring the amount of available retained austenite in crash events. The experimental analysis of microstructural changes during straining was reported mostly for cold-rolled TRIP steels [1,3,4]. Several models describing the kinetics of strain-induced transformation of retained austenite in TRIP steels have been also developed [5]. Haidemenopoulos et al. [6] fitted a mathematical model to available experimental data regarding the evolution of martensite as a function of strain for several TRIP steels. The progressive TRIP effect in hot-rolled sheet steels was investigated rarely [7,8].

Generally, an intensity of the TRIP effect depends strongly on the amount and stability of retained austenite (RA). These steels contain usually 10–15% of γ phase. A chemical composition, an austenite grain size, a type of surrounding phases, and a stress state are the most important factors affecting the stability of retained austenite and the progress of strain-induced transformation. It is well known that carbon content strongly influences the stability of retained austenite. Park et al. [9] observed the correlation between the austenite morphology and carbon content in this phase. They found that the blocky-type RA had lower carbon contents than a film-type RA. Therefore, the blocky RA easily transformed into martensite due to its lower mechanical stability than the film-type RA.

TRIP steels with a ferritic matrix usually contain: ~0.2% C, ~1.5% Mn, and ~1.5% Si or 1.5% Al. Manganese stabilizes the austenitic phase and increases the carbon solubility in ferrite [10]. Sugimoto et al. [11] reported that the addition of 2 wt.% Mn resulted in obtaining a higher fraction of retained austenite. However, the carbon content in this phase became lower. Silicon addition inhibits the carbide precipitation during the bainitic transformation and also strongly increases solid solution strengthening. Silicon can be partially replaced by aluminum due to problems during galvanization, hot-rolling, and welding. However, TRIP steels with Al additions show lower mechanical properties and higher M_s temperatures [12–14]. In order to improve the mechanical properties of such Al-alloyed steels, microadditions of Nb and Ti can be added [8,15–17]. Mo can be also sometimes added to increase strength due to its strong solid solution strengthening effect [18]. Nb and Ti microadditions improve mechanical properties of high-strength low-alloyed (HSLA) steels by precipitation strengthening and grain refinement [8,19–22]. Hausmann et al. [19] reported that the addition of 0.025–0.090 wt.% Nb reduced the cementite precipitation, while the amount of retained austenite increased. Similar results were obtained by Pereloma et al. [20] in 0.2C-1.5Mn-1.5Si-0.039Nb steel.

In the literature, there are several reports concerning the effects of Nb and Ti microadditions on microstructural changes of TRIP steels during straining [1,3,4]. However, most experiments were conducted only under rupture conditions. A limited number of papers addressed the microstructure evolution of thermomechanically processed TRIP steels during interrupted straining. Therefore, the goal of the current work is to characterize microstructural changes of the hot-rolled Si-Al multiphase steel with Nb and Ti microadditions during the interrupted tensile test.

2. Material and Methods

2.1. Material

Investigations concern the Si-Al-Nb-Ti type steel showing a TRIP effect. The detailed chemical composition is listed in Table 1. Silicon was partially replaced by aluminum to improve the manufacturability of the steel sheets. To improve strength, the microadditions of Nb and Ti were added [8,19–22]. The weldability of automotive steel sheets is also a very important issue. Results of studies concerning the weldability of steels with Nb and Ti microadditions showed [23,24] that the fast thermal cycles during welding led to partial dissolution of strengthening phases containing Nb and Ti. As a result of high cooling rate they can precipitate in an uncontrolled way in a heat-affected zone and in the joint. Results of our previous study [25] performed on the 0.24C-1.55Mn-0.87Si-0.4Al-Nb-Ti type steel showed that it is weldable.

Table 1. Chemical composition of the investigated steel, wt.%.

C	Mn	Si	Al	Nb	Ti	P	S	N	O
0.24	1.55	0.87	0.40	0.034	0.023	0.010	0.004	0.0028	0.0006

The investigated steel was prepared by vacuum induction melting. The laboratory ingots were hot-forged to a thickness of 22 mm. Then, they were hot-rolled to a thickness of 4.5 mm in a temperature range 1200–900 °C. Figure 1 shows the applied thermomechanical rolling conditions. It consisted of 3 passes at deformation temperatures: 1050, 950, and 850 °C. The final sheet thickness was ca. 2 mm.

After the final deformation, the flat specimens were air cooled to 700 °C and then more slowly to the temperature of 600 °C within 60 s using a furnace cooling. Then, cooling of the sheets at a rate of about 50 °C/s to the isothermal holding temperature (450 °C) at the bainitic transformation range was applied. The sheet samples were held at 450 °C for 600 s to stabilize retained austenite and finally cooled at a rate of 0.5 °C/s to room temperature.

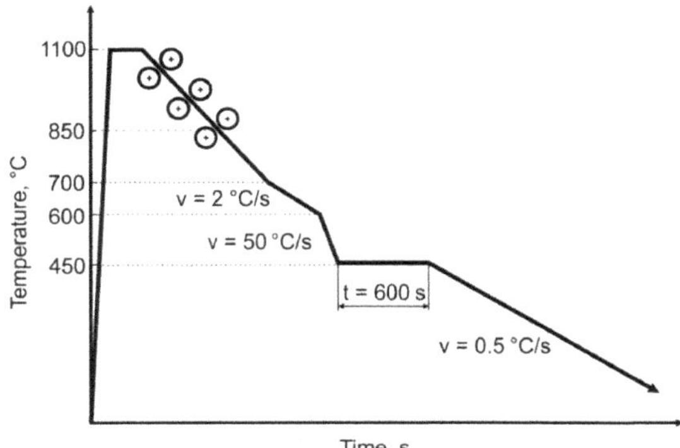

Figure 1. Parameters of thermomechanical processing of investigated steel.

2.2. Description of Experimental Methods

For the detailed investigation of the strain-induced austenite-martensite transformation, interrupted tensile tests were carried out at different strain levels. Samples for the tensile test were cut along the rolling direction from the 2 mm thick sheet. Unidirectional tensile tests were performed to defined strain values of 5%, 10%, 15%, and the final rupture (25%) at a strain rate of 0.008 s^{-1} using a standard tensile test machine Zwick Z/100 (Zwick Roell, Ulm, Germany).

The specimens for optical observations were taken near a fracture area of the deformed specimens according to the tensile direction. The specimens at the initial state (non-deformed) and after the tensile tests were mechanically ground with SiC paper up to 1500 grid and polished using a diamond paste. Then, samples at the initial state were etched in 5% nital and 10% sodium metabisulfite, whereas the deformed samples were etched using the La Pera reagent to observe microstructural details. This type of etching allows obtaining the microstructure in which each phase was characterized by a different color. The 5% nital reagent allows ferrite grains to be revealed, whereas using 10% sodium metabisulfite caused individual microstructural components to be represented by different colors: retained austenite (white), ferrite (gray), bainite (black) and martensite (brown). La Pera reagent allows for color-coded identification of deformed microstructures: ferrite (yellow), retained austenite (light brown), martensite (dark brown). The optical observations were performed using a Zeiss Axio Observer Z1m optical microscope (Carl Zeiss AG, Jena, Germany).

For identifying the amount of individual phases observed in the microstructure, the Image-Pro Plus (version 6.0) software (Media Cybernetics Inc., Rockville, MD, USA) was used. It allows for stereological parameters of individual microstructural constituents to be determined. Based on the optical micrographs, changes in an amount of retained austenite for the specimens deformed at the different strain levels were estimated. Microstructural analysis was performed based on differences in colors of individual phases. For the analysis, optical micrographs were converted to binary maps. A total of 10 digital analyses were performed for each state.

For more detailed investigation of the transformation behavior of retained austenite under applied strain, the tensile specimens were analyzed using an electron backscatter diffraction (EBSD) method. The amount of γ phase was determined based on the average values of 3 measurements. Specimens were prepared using standard metallographic procedures. Then they were electropolished for 40 s, at an operating voltage of 58 V using a TenuPol-5 device (Struers, Ballerup, Denmark) and A8 electrolyte by Struers, (Struers, Ballerup, Denmark) to remove the damage of the surface caused by the grinding and mechanical polishing. The EBSD analyses were conducted at an acceleration voltage of 20 kV and a sample tilt angle of 70° towards normal to an electrooptic beam. EBSD phase maps were recorded using a high resolution scanning electron microscope FEI Inspect F SEM (FEI, Hillsboro, OR, USA) and evaluated by the TSL® OIM software (EDAX OIM Analysis™, NJ, USA). The grain size data were obtained using a grain tolerance angle of 5° and the minimum grain size was chosen to be 2 pixels. All data points with a confidence index (CI) lower than 0.05 were excluded from the analysis.

3. Results

3.1. Initial Microstructure

A microstructure of the investigated steel in the initial state (after hot rolling) is shown in Figure 2. The steel is characterized by a fine-grained multiphase microstructure consisting of ferrite, bainite, and retained austenite. A size of ferritic grains is quite various. Both large and small grains can be observed in the microstructure (Figure 2a). Retained austenite occurs in a form of blocky grains located in a ferritic matrix or as thin layers and small regular grains located inside the bainitic islands.

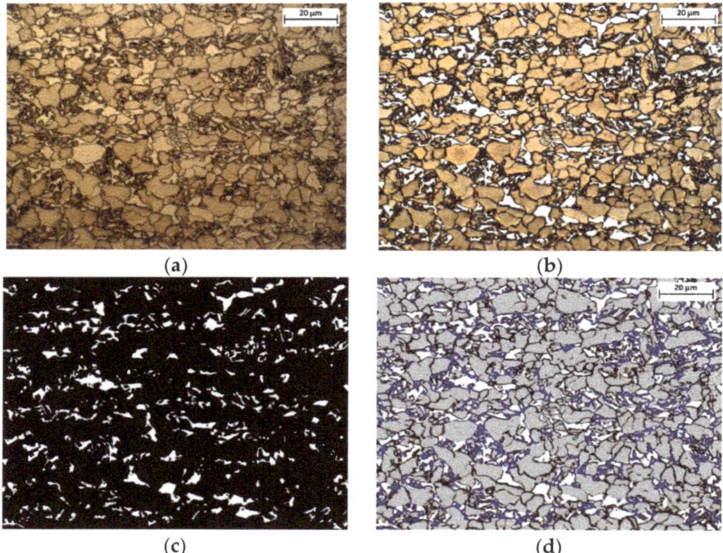

Figure 2. Initial microstructure of the steel: (**a**) Optical micrograph, (**b**) digitally processed optical micrograph, (**c**) binary map of retained austenite, (**d**) combined map: ferrite—gray, retained austenite—white, bainite—blue, grain boundaries—black.

Digital processing of the optical micrograph allowed us to distinguish the fine-grained and layer-type retained austenite (Figure 2b). A grain area of RA was various, in a range from 0.2 to 42 μm^2. Austenitic grains were elongated according to rolling direction (Figure 2c). In Figure 2d individual phases are represented by a different color: ferrite—gray, retained austenite—white, bainite—blue, grain boundaries—black. Figure 3 shows the statistical parameters of retained austenite obtained by the digital processing of optical micrographs. The relatively large quantity of small RA grains can be observed (Figure 3a). Their fraction for a given grain size was estimated to ca. 1–1.5%. However, the largest surface fraction was represented by large grains of RA (Figure 3b). The grains larger than

$10\ \mu m^2$ constitute 30% of the total amount of γ phase (5% of the surface share). The total amount of retained austenite was estimated as 14.6%.

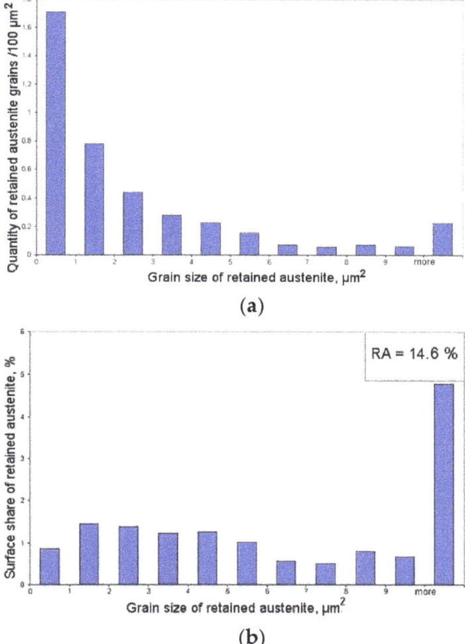

Figure 3. Statistical evaluation of RA parameters: (**a**) average quantity of RA grains per 100 μm^2, (**b**) average surface fraction of RA.

3.2. Interrupted Tensile Test

3.2.1. Sample Deformed to 5% Strain

A specimen deformed to 5% strain during the static tensile test shows the microstructure slightly elongated along to a tensile direction (Figure 4a,b). The largest blocky-type austenitic grains were transformed into martensite (Figure 4c). One can see that some fraction of large austenitic grains transformed only partially—the grain boundaries remained untransformed whereas middle areas of the grains transformed into the martensite (Figure 4d). It can be seen that the amount of small RA grains became higher. It is related to the fact that the newly-formed martensite partially divided the austenitic areas (Figure 4c). As a result of the applied tensile stress, about 3% of martensite, characterized by various sizes, was formed in the microstructure (Figure 4c).

Figure 5 shows the statistical parameters of the retained austenite and martensite obtained by the digital processing of optical micrographs of the specimen deformed to 5% strain. For the quantitative analysis of the RA islands which remain stable after deformation, it was assumed that if the martensitic transformation took place (even only partially) in an austenite grain, such a grain was identified as transformed. The amount of retained austenite in deformed specimens was compared to the microstructure at the initial state (100%) in order to assess an amount of austenite transformed into martensite during straining.

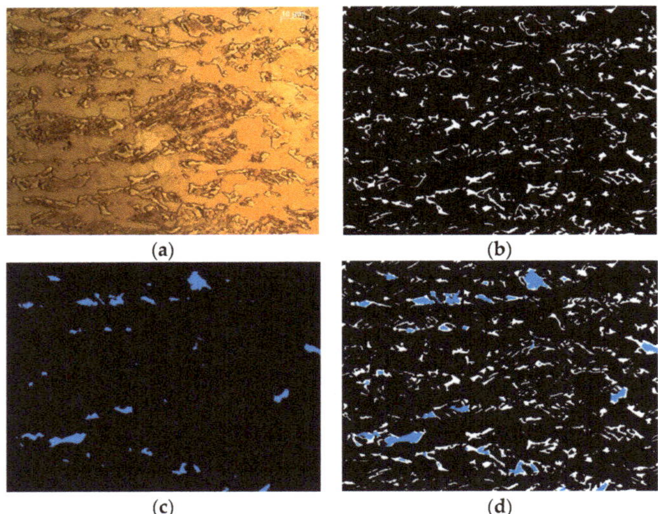

Figure 4. Microstructure of the steel deformed to 5% strain: (**a**) optical micrograph, (**b**) binary map of retained austenite, (**c**) binary map of martensite, (**d**) binary map of retained austenite (white) and martensite (blue).

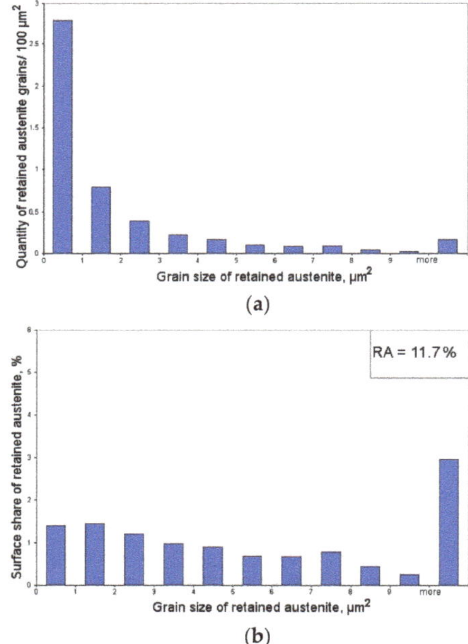

Figure 5. Statistical evaluation of RA parameters at 5% strain: (**a**) average quantity of RA grains per 100 μm^2, (**b**) average surface fraction of RA.

The quantity and surface fraction of the smallest austenitic grains became higher when compared to the specimen at the initial state (Figure 5). However, the surface fraction of the largest RA grains remained significantly higher (Figure 5b). An amount of retained austenite observed in the microstructure of the specimen deformed to 5% strain was estimated to be 11.7%. This means that 2.8% of the total amount of γ phase transformed into martensite (Figure 5b). Figure 6 shows that 70% of the fraction of austenitic grains larger than 10 μm^2 transformed into the strain-induced martensite. Smaller grains of RA showed the lower tendency to strain-induced transformation. Austenitic grains smaller than 3 μm^2 remained almost stable (Figure 6).

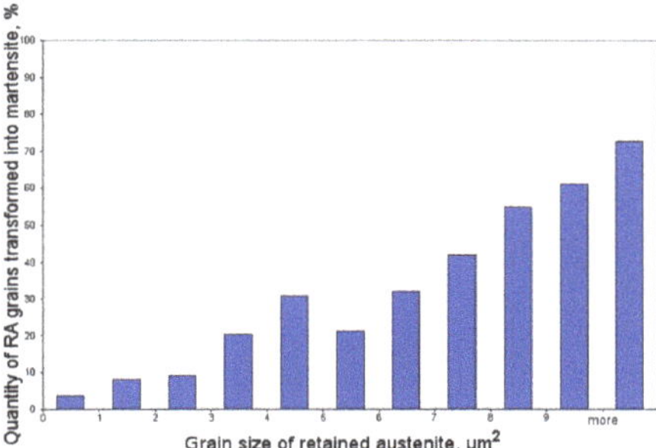

Figure 6. Quantity of retained austenite grains transformed into martensite for the specimen deformed to 5% strain.

3.2.2. Sample Deformed to 10% Strain

A sample deformed to 10% strain is characterized by grains elongated according to a tensile direction (Figure 7). The amount of RA grains larger than 4 μm^2 significantly decreased (Figure 8). Almost all austenitic grains larger than 7 μm^2 transformed into martensite. The central areas of the large austenite grains transformed into martensite, while the regions located near the grain boundaries remained stable (Figure 7d). As a result of microstructure fragmentation, a quantity of small grains of γ phase increased (Figure 8a). A large amount of small grains (<1 μm^2) and layers of retained austenite located at bainitic islands remained unchanged. The quantity of small martensitic areas was high. However, the greatest surface area showed the largest martensitic islands (Figure 7c).

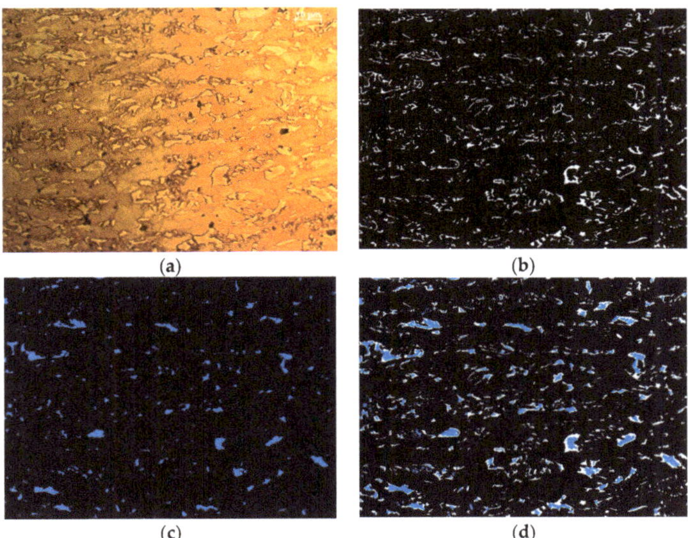

Figure 7. Microstructure of investigated steel deformed to 10% strain: (**a**) optical micrograph, (**b**) binary map of retained austenite, (**c**) binary map of martensite, (**d**) binary map of retained austenite (white) and martensite (blue).

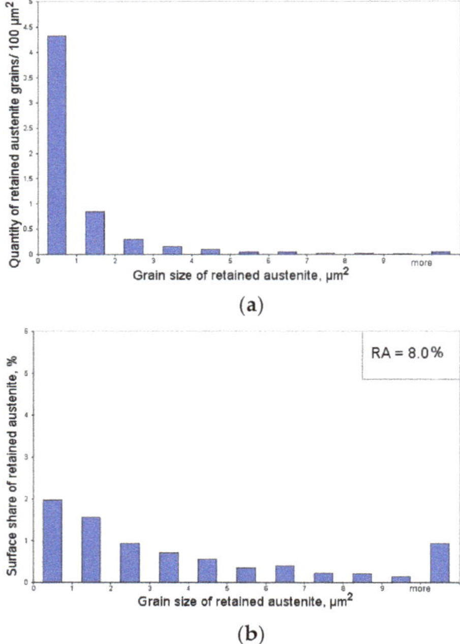

(a)

(b)

Figure 8. Statistical evaluation of RA parameters at 10% strain: (**a**) average quantity of RA grains per 100 μm², (**b**) average surface share of RA.

Figure 8 shows the statistical parameters of retained austenite and martensite obtained by digital processing of 10% strain optical micrographs. Austenitic grains in a size range of 1–7 μm² were transformed at different proportions (Figure 8). Statistical calculations showed that small austenitic grains were dominant in the microstructure of 10% strain specimen. Both the quantity and surface share of small RA grains (Figure 8a) were greater than for the large grains (Figure 8b). In case of the specimen at the initial state, the highest surface share showed austenitic grains larger than 10 μm² (Figure 3b). The amount of retained austenite detected for the specimen deformed to 10% strain was estimated to about 8%. It means that ~6.6% of the total RA amount transformed into the martensite.

Figure 9 shows that austenitic grains smaller than 1 μm² remained mechanically stable. Only 15% of the smallest grains transformed into martensite. It is worth to note that the largest grains, which only partially transformed into martensite, were also counted as the transformed grains (Figure 7d). It resulted in a quite inflated amount of calculated γ phase (Figure 9).

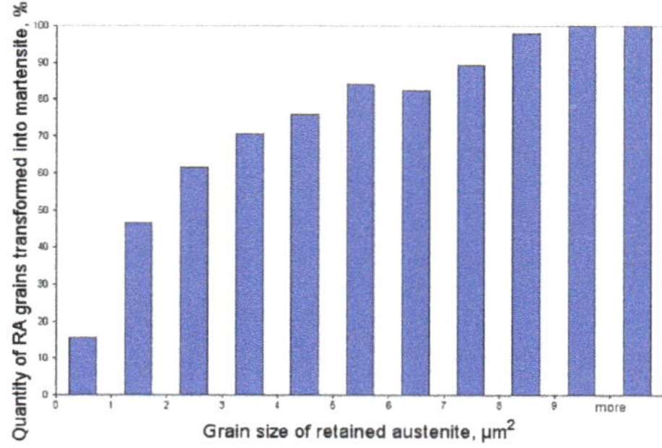

Figure 9. Quantity of retained austenite grains transformed into martensite for the specimen deformed to 10% strain.

3.2.3. Sample Deformed to 15% Strain and to the Rupture

Microstructural components of the 15% strained sample were elongated according to a tensile stress direction (Figure 10a). All blocky grains of retained austenite located in a ferritic matrix underwent the martensitic transformation. Large austenitic grains located at the bainitic islands almost completely transformed into the martensite. Thin RA layers only partially changed into martensite (Figure 10a). Amount of small austenitic grains became higher due to the fragmentation of large austenite grains by the newly-formed martensite. Amount of retained austenite was estimated to about 6%. It means that more than a half fraction of retained austenite transformed into the martensite during straining (Figure 3b). A surface share of martensite was ca. 10%. In the deformation range from 10% to 15%, about 2% of retained austenite transformed into the martensite, which corresponds to ca. 14% of the total share of the γ phase.

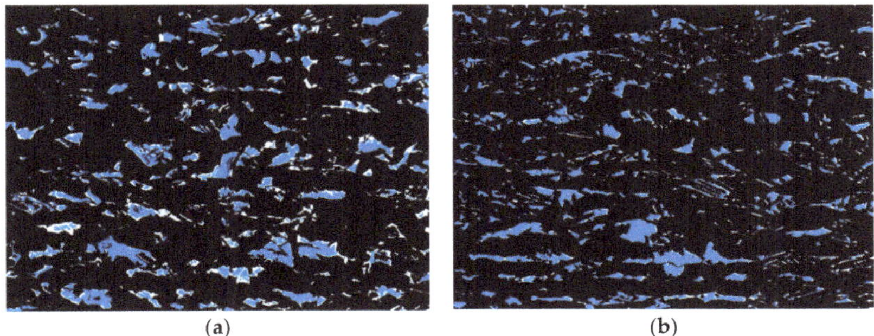

(a)   (b)

Figure 10. Binary map of retained austenite (white) and martensite (blue) for the specimens deformed to: (**a**) 15% strain, (**b**) 25% strain (a rupture of the sample).

Figure 10b shows the sample deformed to 25% strain (to rupture). The microstructure was strongly deformed along a tensile load. Several small, thin layers located at bainitic islands remained stable. Only the smallest grains and some grain boundaries of RA did not transform into martensite (Figure 10b). The largest martensitic areas were formed from the largest grains of γ phase. The amount of retained austenite remained in the microstructure of the sample deformed to rupture is ca. 2.5%. The martensitic fraction was estimated to ~11.6%. In the deformation range from 15% to 25%, about 3.5% of the retained austenite transformed into strain-induced martensite, which corresponds to about 25% of the total share of γ phase.

3.3. Electron Backscatter Diffraction (EBSD) Results

The microstructure of the investigated steel was also examined by using an EBSD method. The amount of RA at the initial state was 13.8% [15]. Figure 11a–c provides the selected EBSD maps of the steel deformed to 5% strain. Figure 11a shows the IQ (image quality) map. Ferrite is represented by the brightest regions characterized by the best diffraction quality. The retained austenite, bainite, and grain boundaries are represented by different grey levels. Martensite occurred as the darkest areas characterized by lowest values of the IQ parameter. The fraction of low-angle boundaries (misorientation < 15°) marked as red and green lines was estimated to be 25% (Figure 11b). High-angle boundaries marked as blue and orange lines (misorientation > 15°) were dominant. The strain-induced martensitic transformation occurred inside the largest austenite grains located near ferrite grains (Figure 11c). The central areas of the large austenite grains transformed into martensite, whereas the regions located near the grain boundaries remained stable. Retained austenite in form of small grains and thin layers located at bainitic islands remained unchanged (Figure 11c). The average amount of RA was estimated to about 10.1%.

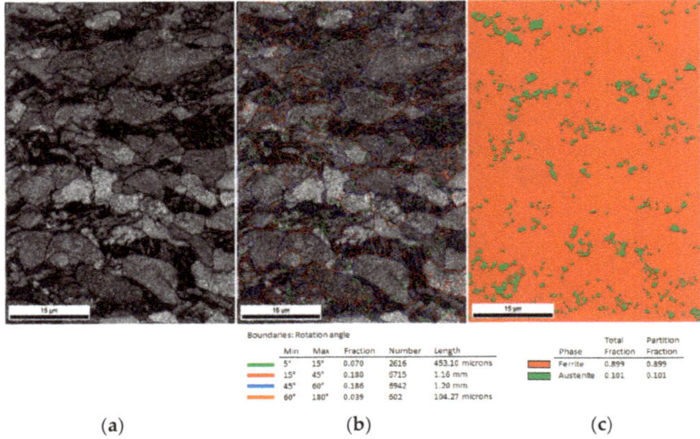

Figure 11. EBSD analysis of steel deformed to 5% strain: (**a**) image quality (IQ) map, (**b**) image quality (IQ) map displaying the boundary character, (**c**) phase distribution map—retained austenite as green.

Figure 12a shows the IQ map of steel deformed to 10% strain. The amount of regions represented by the lowest IQ parameter is slightly higher when compared to the specimen deformed to 5% strain (Figure 11a). It is related to the increase in the amount of martensite, which was characterized by the lowest value of the IQ parameter. The fraction of low-angle boundaries was estimated to be about 50% (Figure 12b). The amount of high-angle boundaries was lower when compared to the specimen deformed to 5% strain (Figure 11b). Dislocations generated during plastic deformation were reflected as an increase in the number of low-angle boundaries. On the other hand, the newly-formed martensite-austenite boundaries, as a result of strain-induced transformation, generated new high-angle boundaries. Therefore, the final distribution between low-angle and high-angle boundaries was a synergistic result of these two effects. The highest stability showed austenitic grains located at bainitic islands. However, some fraction of RA at these areas transformed into martensite (Figure 12c). Further fragmentation of larger RA grains was also observed. The amount of γ phase remained stable was estimated to about 7.7%. Figure 13a shows the IQ map of steel deformed to 25% strain (a rupture of the sample). An increase in the amount of areas characterized by the lowest IQ value was observed when compared to the samples deformed to 5% and 10% strain. The amount of low and high-angle boundaries was similar (Figure 13b). Austenite grains are strongly fragmented. They were located mainly at bainitic areas (Figure 13c). The amount of RA was estimated to 3.7%.

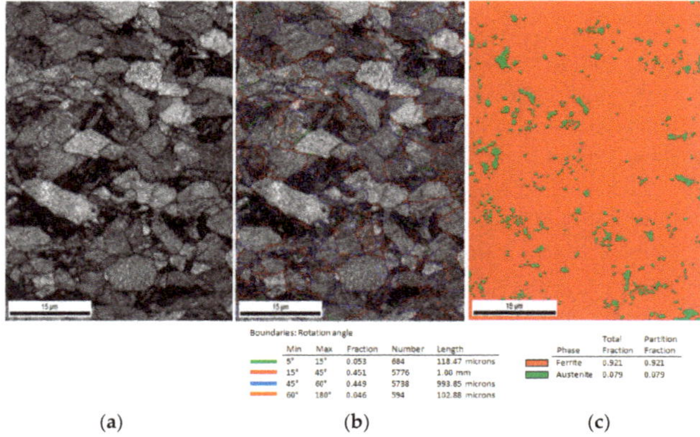

Figure 12. EBSD analysis of steel deformed to 10% strain: (**a**) image quality (IQ) map, (**b**) image quality (IQ) map displaying the boundary character, (**c**) phase distribution map—retained austenite as green.

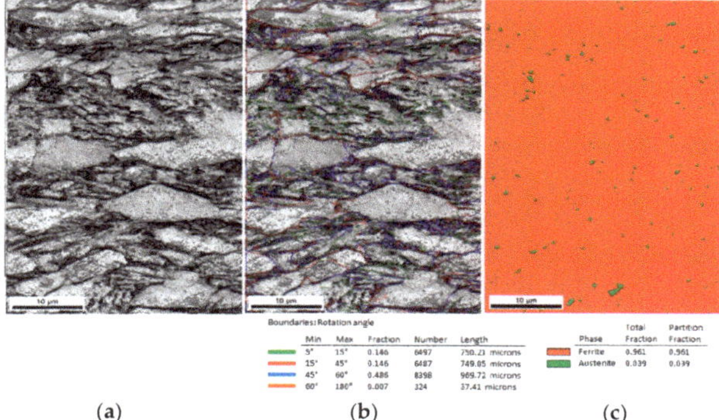

Boundaries: Rotation angle

	Min	Max	Fraction	Number	Length
▬	5°	15°	0.146	6497	750.23 microns
▬	15°	45°	0.146	6487	749.05 microns
▬	45°	60°	0.486	8398	969.72 microns
▬	60°	180°	0.007	324	37.41 microns

	Phase	Total Fraction	Partition Fraction
▬	Ferrite	0.961	0.961
▬	Austenite	0.039	0.039

(a)   (b)   (c)

Figure 13. EBSD analysis of steel deformed to 25% strain (a rupture of the sample): (**a**) image quality (IQ) map, (**b**) image quality (IQ) map displaying the boundary character, (**c**) phase distribution map—retained austenite as green.

4. Discussion

A volume fraction and stability of retained austenite are essential factors in designing TRIP steels. Numerous reports concern the stability of retained austenite in TRIP-assisted multiphase steels [3,4,6,26,27]. Nowadays, researchers are focused on C-Mn-Si-Al TRIP steels with microadditions such as Nb, Ti, and Mo [4,8,17–21,28]. It is related to the fact, that these microadditions improve the properties by enhancing the strength and grain refinement [8–19,21]. It is well known that the amount of retained austenite was decreasing as a result of plastic deformation [1–9]. However, the kinetics of TRIP effect is a complex issue and it depends on several factors (i.e., chemical composition, morphology, and grain size of RA). It is especially important to monitor the microstructure evolution during straining due to the detailed characterization of the tendency of individual RA grains to martensitic transformation [29]. However, most scientific reports concern the comparison of microstructural features of a specimen at an initial state (non-deformed) and deformed to rupture [7]. Moreover, the effects of grain size and morphology on the stability of retained austenite in hot-rolled TRIP grades were analyzed rarely. Therefore, this problem was discussed in the present study. A quantitative microstructural analysis of samples deformed gradually in interrupted tensile tests, assessed by using both digital image analysis and EBSD methods, allowed the tendency of RA for martensitic transformation, depending on grain type and its size, to be monitored.

An amount of retained austenite decreased as the strain level increased (Figure 14). The highest relative fraction of γ phase transformed at the lowest strain level (5%). In this case, about 90% of austenitic grains larger than 10 μm^2 transformed into martensite. As the deformation level increases, smaller grains of austenite further transformed. For the deformation level 10% the austenitic grains bigger than 8 μm^2 transformed into martensite. It is related to the fact that large blocky-type grains of retained austenite were characterized by relatively low mechanical stability due to a lower carbon content. Moreover, blocky-type austenite possesses some fraction of microstructural defects, like dislocations and stacking faults which constitute the martensite nucleation zones [30]. Some reports [31–33] showed that film-type of retained austenite located at the bainitic islands is more beneficial due to the higher strength and steel toughness. However, Xu et al. [34] reported that the most favorable, gradual progress of martensitic transformation occurred for the microstructure consisting of both blocky and film types of retained austenite. A neighborhood of soft ferrite also favors the easier transformation into martensite [35]. Carbon content strongly affects the austenite stabilization (decreases the M_s temperature). Retained austenite characterized by low carbon content can easily be transformed into martensite during plastic deformation. High carbon content in the austenite caused its excessive stabilization, which also resulted in ductility reduction [36]. Pereloma et al. [37] reported that the most favorable, gradual TRIP effect occurs when retained austenite contains 1.1–1.6% C. The detailed information concerning the X-ray diffraction analysis of investigated steel, including the estimation of carbon content in the retained austenite, can be found in our earlier work [38]. The determined carbon content was 1.28 wt.%, giving the opportunity for the gradual transformation. Increasing the strain level to 15% resulted in the complete transformation of austenitic grains larger than 6 μm^2. In the case of the specimen deformed to 25% strain (to rupture), austenitic grains larger than 2 μm^2 transformed into martensite (Figure 13). Only thin films and small grains of RA located at the bainitic islands remained stable. As the grain size of RA decreased, its mechanical stability increased. The new-formed grain boundaries acted as obstacles to new potential martensitic laths, resulting in inhibiting a further transformation [39].

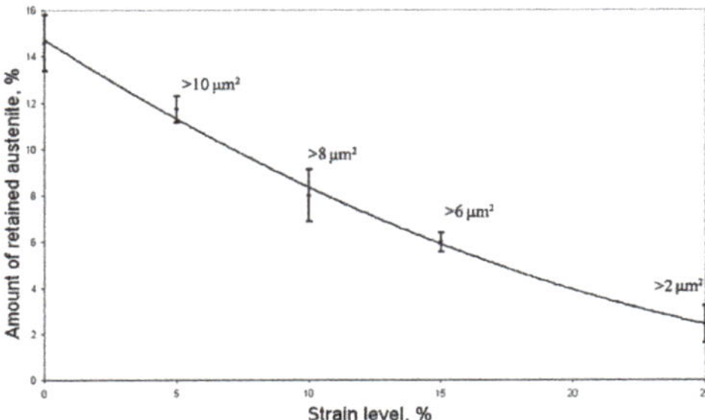

Figure 14. Change in the amount of retained austenite for defined grain size as a function of strain level.

An amount of retained austenite estimated by using digital processing of optical micrographs and EBDS were similar (Table 2). However, the amount of RA estimated by using EBSD method was slightly lower when compared to the image analysis. It is related to the fact that the image analysis contained some uncertainty of measurement. Digital processing of optical micrographs was characterized by lower resolution than EBSD method. Both applied methods confirmed that the large grains of RA were more prone to martensitic transformation than small grains and thin layers of this phase. Moreover, it was clearly seen, that γ phase located near grain boundaries remained stable due to a higher carbon content. Similar results were obtained by Park et al. [9] in 0.2C-2Mn-1Si steel. Additionally, they also observed that the amount of RA decreased after applying 15% strain; after that, the intensity of the strain-induced martensitic transformation reduced. Figure 15 shows that the surface share of austenitic grains larger than 10 μm^2 decreased as the deformation level increased. The mount of grains in the size range from 3 to 10 μm^2 was decreasing slowly. The amount of the smallest austenitic grains (<1 μm^2) was increasing, and the deformation level increased. It is related to the fragmentation of large austenite grains by newly-formed martensite (Figure 15).

Table 2. Change in the amount of retained austenite as a function of strain level estimated by using image analysis and EBSD method.

Strain Level, %	Amount of RA (Image Analysis),%	Standard Deviation, %	Amount of RA (EBSD Method), %
0	14.6	1.3	13.8
5	11.7	1.2	10.9
10	8.0	1.2	7.9
15	6.0	1.0	5.8
25	2.4	0.8	3.9

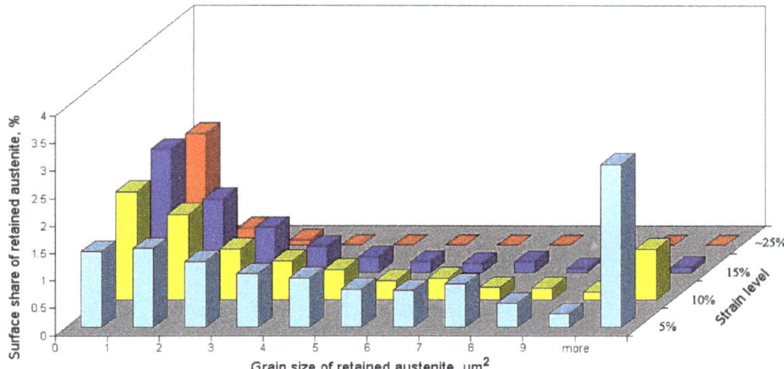

Figure 15. Distribution of grain size and the surface share of retained austenite as a function of strain level.

A correlation between grain size of retained austenite and its tendency for martensitic transformation as a function of strain level is shown in Figure 16. For the deformation level 5%, mostly large austenitic grains transformed into martensite. The tendency can be approximated by a linear function. For the higher deformation levels, the tendency was approximated by a parabolic function (Figure 16). As the deformation level increased, the intensity of martensitic transformation decreased, which was related to the lower amount of large austenite grains. A total of 30% of the smallest austenitic grains and layers transformed into martensite (Figure 16). These areas were characterized by the highest mechanical stability.

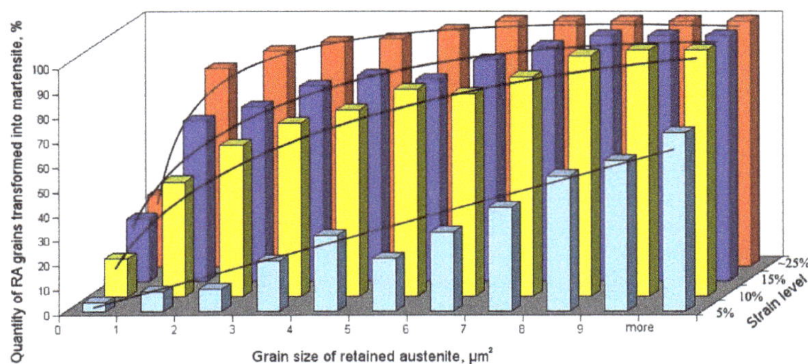

Figure 16. Correlation between grain size of retained austenite and its tendency to martensitic transformation as a function of strain level.

5. Conclusions

The present study addressed the quantitative analysis of the microstructure evolution in thermomechanically-processed Nb-Ti-microalloyed Si-Al multiphase steel subjected to interrupted tensile tests. The main findings of the paper can be summarized as follows:

- Martensitic transformation began in the central area of large blocky-type austenitic grains located in the ferritic matrix. As the deformation level increased the intensity of martensitic transformation decreased due to the lower amount of large austenite grains.
- A size of austenite grains which transformed into martensite decreased when deformation level was increasing.

- An amount of small austenitic grains increased along with the deformation level due to the fragmentation of large austenitic grains by newly-formed martensite. The corresponding fractions of low-angle and high-angle boundaries were a synergistic effect of the dislocation increase due to straining and new boundaries formed by strain-induced martensite formation.
- Austenitic grains smaller than 1 μm^2 and thin layers showed the highest mechanical stability. It was due to their high mechanical stability related to the relatively high carbon content.
- An amount of retained austenite detected in the specimen at the initial state was ca. 14%. Specimens deformed up to rupture possessed ca. 2.5% of RA (i.e., ~17% of untransformed retained austenite).

Author Contributions: Conceptualization, A.G.; data curation, A.G. and A.K.; formal analysis, A.G.; funding acquisition, A.G.; investigation, A.K. and K.R.; methodology, A.G., A.K. and K.R.; project administration, A.G.; validation, A.S.; visualization, A.S.; writing—original draft, A.K.; writing—review and editing, K.R. and A.S.

Funding: The financial support of the National Science Center, Poland, is gratefully acknowledged, grant no. 2017/27/B/ST8/02864.

Conflicts of Interest: The authors declare no conflict of interest.

References

1. Streicher-Clarke, A.M.; Speer, J.G.; Matlock, D.K.; De Cooman, B.C.; Williamson, D.L. Analysis of lattice parameter changes following deformation of a 0.19C-1.63Si-1.59Mn Transformation-Induced Plasticity sheet steel. *Metall. Mater. Trans. A* **2005**, *36*, 907–918. [CrossRef]
2. Suliga, M.; Muskalski, Z.; Wiewiorowska, S. The influence of drawing speed on fatigue strength TRIP steel wires. *Arch. Civ. Mech. Eng.* **2009**, *9*, 97–107. [CrossRef]
3. Kobayashi, J.; Tonegawa, H.; Sugimoto, K. Cold formability of 22SiMnCrB TRIP-aided martensitic sheet steel. *Proc. Eng.* **2014**, *81*, 1336–1341. [CrossRef]
4. Krizan, D.; De Cooman, B.C. Analysis of the strain-induced martensitic transformation of retained austenite in cold rolled micro-alloyed TRIP steel. *Steel Res. Int.* **2008**, *79*, 513–522. [CrossRef]
5. Samek, L.; De Moor, E.; Penning, J.; De Cooman, B.C. Influence of alloying elements on the kinetics of strain-induced martensitic nucleation in low-alloy, multiphase high-strength steels. *Metall. Mater. Trans. A* **2006**, *37*, 109–124. [CrossRef]
6. Haidemenopoulos, G.N.; Aravas, N.; Bellas, I. Kinetics of strain-induced transformation of dispersed austenite in low-alloy TRIP steels. *Mater. Sci. Eng. A* **2014**, *615*, 416–423. [CrossRef]
7. Kaar, S.; Krizan, D.; Schwabe, J.; Hofmann, H.; Hebesberger, T.; Commenda, C.; Samek, L. Influence of the Al and Mn content on the structure-property relationship in density reduced TRIP-assisted sheet steels. *Mater. Sci. Eng. A* **2018**, *735*, 475–486. [CrossRef]
8. Kucerova, L.; Bystriansky, M. Comparison of thermo-mechanical treatment of C-Mn-Si-Nb and C-Mn-Si-Al-Nb TRIP steels. *Proc. Eng.* **2017**, *207*, 1856–1861. [CrossRef]
9. Park, H.S.; Han, J.C.; Lim, N.S.; Seol, J.B.; Park, C.G. Nano-scale observation on the transformation behavior and mechanical stability of individual retained austenite in CMnSiAl TRIP steels. *Mater. Sci. Eng. A* **2015**, *627*, 262–269. [CrossRef]
10. Jabłońska, M.B.; Śmiglewicz, A.; Niewielski, G. The effect of strain rate on the mechanical properties and microstructure of the high-Mn steel after dynamic deformation tests. *Arch. Metall. Mater.* **2015**, *60*, 577–580. [CrossRef]
11. Sugimoto, K.; Usui, N.; Kobayashi, M.; Hashimoto, S. Effects of volume fraction and stability of retained austenite on ductility of TRIP-aided dual-phase steels. *ISIJ Int.* **1992**, *32*, 1311–1318. [CrossRef]
12. De Cooman, B.C. Structure—Properties relationship in TRIP steels containing carbide-free bainite. *Solid State Mater. Sci.* **2004**, *8*, 285–303. [CrossRef]
13. Jacques, P.J.; Girault, E.; Catlin, T.; Geerlofs, N.; Kop, T.; Van der Zwaag, S.; Delannay, F. Bainite transformation of low carbon Mn-Si TRIP-assisted multiphase steels: Influence of silicon content on cementite precipitation and austenite retention. *Mater. Sci. Eng. A* **1999**, *273*, 475–479. [CrossRef]
14. Girault, E.; Mertens, A.; Jacques, P.; Hubaert, Y.; Verlinden, B.; Van Humbeeck, J. Comparison of the effect of silicon and aluminium on the tensile behavior of multiphase TRIP-assisted steels. *Scr. Mater.* **2001**, *44*, 885–892. [CrossRef]

15. Grajcar, A.; Radwanski, K. Microstructural comparison of the thermomechanically treated and cold deformed Nb-microalloyed TRIP steel. *Mater. Tehnol.* **2014**, *48*, 679–683.

16. Javaheria, V.; Khodaie, N.; Kaijalainen, A.; Porter, D. Effect of niobium and phase transformation temperature on the microstructure and texture of a novel 0.40% C thermomechanically processed steel. *Mater. Charact.* **2018**, *124*, 295–308. [CrossRef]

17. Opiela, M.; Grajcar, A. Hot deformation behavior and softening kinetics of Ti-V-B microalloyed steels. *Arch. Civ. Mech. Eng.* **2012**, *12*, 327–333. [CrossRef]

18. Wang, C.; Ding, H.; Tang, Z.Y.; Zhang, J. Effect of isothermal bainitic processing on microstructures and mechanical properties of novel Mo and Nb microalloyed TRIP steel. *Ironmak. Steelmak.* **2015**, *42*, 9–16. [CrossRef]

19. Hausmann, K.; Krizan, D.; Spiradek-Hahn, K.; Pichler, A.; Werner, E. The influence of Nb on transformation behavior and mechanical properties of TRIP-assisted bainitic–ferritic sheet steels. *Mater. Sci. Eng. A* **2013**, *588*, 142–150. [CrossRef]

20. Pereloma, E.V.; Timokhina, I.B.; Hodgson, P.D. Transformation behaviour in thermomechanically processed C-Mn-Si TRIP steels with and without Nb. *Mater. Sci. Eng. A* **1999**, *273–275*, 448–452. [CrossRef]

21. Kammouni, A.; Saikaly, W.; Dumont, M.; Marteaud, C.; Banod, X.; Charai, A. Effect of the bainitic transformation temperature on retained austenite fraction and stability in Ti microalloyed TRIP steels. *Mater. Sci. Eng. A* **2009**, *518*, 89–96. [CrossRef]

22. Sozańska-Jędrasik, L.; Mazurkiewicz, J.; Borek, W.; Matus, K. Carbides analysis of the high strength and low density Fe-Mn-Al-Si steels. *Arch. Metall. Mater.* **2018**, *63*, 265–276.

23. Gorka, J.; Opiela, M. Structure and properties of high-strength low-alloy steel melted by the laser beam. *Mater. Perform. Charact.* **2019**, *8*, 1–10. [CrossRef]

24. Gorka, J.; Janicki, D.; Fidali, M.; Jamrozik, W. Thermographic assessment of the HAZ properties and structure of thermomechanically treated steel. *Int. J. Thermophys.* **2017**, *38*, 183. [CrossRef]

25. Grajcar, A.; Grzegorczyk, B.; Różański, M.; Stano, S. Microstructural aspects of bifocal laser welding of TRIP steels. *Arch. Metall. Mater.* **2017**, *62*, 611–618. [CrossRef]

26. Tang, T.Y.; Huang, J.N.; Ding, H.; Cai, Z.H.; Misra, R.D.K. Austenite stability and mechanical properties of a low-alloyed ECAPed TRIP-aided steel. *Mater. Sci. Eng. A* **2018**, *724*, 95–102. [CrossRef]

27. Wang, X.D.; Huang, B.X.; Rong, Y.H.; Wang, L. Microstructures and stability of retained austenite in TRIP steels. *Mater. Sci. Eng. A* **2006**, *438–440*, 300–305. [CrossRef]

28. Tang, Z.; Ding, H.; Ding, H.; Cal, M.; Du, L. Effect of prestrain on microstructures and properties of Si-Al-Mn TRIP steel sheet with niobium. *J. Iron Steel Res. Int.* **2010**, *17*, 59–65. [CrossRef]

29. Zhang, S.; Findley, K.O. Quantitative assessment of the effects of microstructure on the stability of retained austenite in TRIP steels. *Acta Mater.* **2013**, *61*, 1895–1903. [CrossRef]

30. Das, A.; Ghosh, M.; Tarafder, S.; Sivaprasad, S.; Chakrabarti, D. Micromechanisms of deformation in dual phase steels at high strain rates. *Mater. Sci. Eng. A* **2017**, *680*, 249–258. [CrossRef]

31. Gao, G.; Zhang, B.; Cheng, C.; Zhao, P.; Zhang, H.; Bai, B. Very high cycle fatigue behaviors of bainite/martensite multiphase steel treated by quenching-partitioning-tempering process. *Int. J. Fatigue* **2016**, *92*, 203–210. [CrossRef]

32. Shen, Y.F.; Qiu, L.N.; Sun, X.; Zuo, L.; Liaw, P.K.; Raabe, D. Effects of retained austenite volume fraction, morphology, and carbon content on strength and ductility of nanostructured TRIP-assisted steels. *Mater. Sci. Eng. A* **2015**, *636*, 551–564. [CrossRef]

33. Zhou, Q.; Qian, L.; Tan, J.; Meng, J.; Zhang, F. Inconsistent effects of mechanical stability of retained austenite on ductility and toughness of transformation-induced plasticity steels. *Mater. Sci. Eng. A* **2013**, *578*, 370–376. [CrossRef]

34. Xu, Y.; Hu, Z.; Zou, Y.; Tan, X.; Han, D.; Chen, S.; Ma, D.; Misra, R.D.K. Effect of two-step intercritical annealing on microstructure and mechanical properties of hot-rolled medium manganese TRIP steel containing δ-ferrite. *Mater. Sci. Eng. A* **2017**, *688*, 40–55. [CrossRef]

35. Sugimoto, K.; Misu, M.; Kobayashi, M.; Shirasawa, H. Effects of second phase morphology on retained austenite morphology and tensile properties in a TRIP-aided dual phase steel sheet. *ISIJ Int.* **1993**, *33*, 775–782. [CrossRef]

36. Timokhina, I.; Hodgson, P.; Pereloma, E.V. Effect of microstructure on the stability of retained austenite in transformation-induced-plasticity steels. *Metall. Mater. Trans. A* **2004**, *35*, 2331–2341. [CrossRef]

37. Pereloma, E.V.; Gazder, A.A.; Timokhina, I.B. Retained austenite: Transformation-Induced Plasticity. In *Encyclopedia of Iron, Steel, and Their Alloys*; Taylor and Francis: New York, NY, USA, 2016; pp. 3088–3103. [CrossRef]

38. Grajcar, A.; Krztoń, H. Effect of isothermal bainitic transformation temperature on retained austenite fraction in C-Mn-Si-Al-Nb-Ti TRIP-type steel. *J. Achiev. Mater. Manuf. Eng.* **2009**, *35*, 169–176.

39. Jimenez-Melero, E.; Van Dijk, N.; Zhao, L.; Sietsma, J.; Offerman, S. Characterization of individual retained austenite grains and their stability in low-alloyed TRIP steels. *Scr. Mater.* **2007**, *55*, 6713–6723. [CrossRef]

Article

Evolution of Microstructure and Hardness of High Carbon Steel under Different Compressive Strain Rates

Rumana Hossain *, Farshid Pahlevani and Veena Sahajwalla

Centre for Sustainable Materials Research and Technology (SMaRT@UNSW), School of Materials Science and Engineering, The University of New South Wales (UNSW Sydney), Sydney, NSW 2052, Australia; f.pahlevani@unsw.edu.au (F.P.); veena@unsw.edu.au (V.S.)
* Correspondence: r.hossain@unsw.edu.au; Tel.: +61-405-408-810

Received: 20 June 2018; Accepted: 24 July 2018; Published: 26 July 2018

Abstract: Understanding the effect of high strain rate deformation on microstructure and mechanical property of metal is important for addressing its performance as high strength material. Strongly motivated by the vast industrial application potential of metals having excellent hardness, we explored the phase stability, microstructure and mechanical performance of an industrial grade high carbon steel under different compressive strain rates. Although low alloyed high carbon steel is well known for their high hardness, unfortunately, their deformation behavior, performance and microstructural evolution under different compressive strain rates are not well understood. For the first time, our investigation revealed that different strain rates transform the metastable austenite into martensite at different volume, simultaneously activate multiple micromechanisms, i.e., dislocation defects, nanotwining, etc. that enhanced the phase stability and refined the microstructure, which is the key for the observed leap in hardness. The combination of phase transformation, grain refinement, increased dislocation density, formation of nanotwin and strain hardening led to an increase in the hardness of high carbon steel.

Keywords: strain-induced transformation; microstructure; mechanical property; high carbon steel

1. Introduction

Multiphase high carbon steels with metastable retained austenite (RA) phase are excellent for industrial applications due to their high hardness and abrasion resistance. When the metastable RA in high carbon steel is subjected to harsh environmental conditions; i.e., compression [1,2], abrasion [3], impact [4], etc.; RA transforms to strain induced α' and ε martensite [5]. The consequence of this transformation is dependent on various parameters, such as morphology [6], chemical composition [1,6,7] and stacking fault energy of austenite [8]. The α' martensite is found to be nucleated at the intersection of the shear bands, while the ε-martensites are formed due to the overlapping stacking faults [8]. These strain-induced martensites possess more strength and hardness than austenite resulting in increasing the strain hardening ability and strengthening of the material [9–12].

Ample amount of studies have been proposed to describe the transformation induced plasticity effect in austenitic steel [1,13]. Previous studies have demonstrated that the hardness of steel increases with decreasing grain size, which is known as the Hall-Petch effect [14]. In this effect, the strengthening originates from the fact that grain boundaries block the dislocation movement; hence, the material has more resistance to deformation. The high strain rate induced nanotwin boundaries in the face-centered cubic structure also act as grain boundaries and block the dislocation movements, therefore, enhance the structural strength [15].

Strain rates are considered to have a significant effect on the deformation behavior of steels [16,17]. Superplastic behavior of 0.9% C steel at a low strain rate (up to 10^{-3}/s) was assessed in one study, showed positive strain rate sensitivity of the material. Another study shows the formation of adiabatic shear in 0.77% carbon steel when subjected to high strain rate compression testing conducted at various regimes of temperature [18]. Limited studies have been carrying out on mechanical performance evaluation at various strain rate deformation and its relation with the microstructure of high carbon steel. The hindrance behind conducting mechanical tests on high carbon steel is their exceptional hardness and strength accompanied by machine capabilities, the brittle nature and hence the absence of interest for their mechanical behavior at varying strain rates. However, the study of the mechanical performance of deformation-induced high carbon steels are very important as they are broadly used in a various industrial application where severe impacts and compression are encountered, and high abrasion resistance is required. We carefully chose the compressive stress lower than the fracture stress to avoid fracture or failure of the material. This will help us to demonstrate the mechanical and microstructural behavior of steel at the extreme working condition.

This study focused on dual-phase high carbon steel with martensite and RA phases for the use in extreme operating conditions, i.e., wear, abrasion, impact, compression, etc. Its microstructural behavior at different strains has been investigated by X-ray diffraction, electron backscattering diffraction and transmission electron microscopy. In addition, the strain hardening effect on the overall hardness of the high carbon steel was investigated by nanohardness analyses. Identifying various micromechanisms and their effect on hardness at various strain rates is essential to characterize high carbon steel as a superior material for the different industrial application.

2. Experimental

In this study, duplex martensitic high carbon steel having ~45% of RA with the chemical composition of 1.0%C-0.2%Si-1.0%Mn-0.6%Cr (in wt %), was investigated. Standardized compression tests were performed in a computer-controlled servo-hydraulic uniaxial testing machine (Instron 8510). The tests were conducted for five different strain rates of 2×10^{-5}/s, 2×10^{-4}/s, 2×10^{-3}/s, 2×10^{-2}/s and 2×10^{-1}/s at room temperature and the load induced until the compressive stress reached 2000 MPa for each sample with the size of 4 mm $\times$ 4 mm $\times$ 4 mm. We have chosen this compressive stress to understand the hardness and properties of this steel under extreme stress but before failure. Five samples were compressed for each experimental condition for statistical data reliability. Standard metallographic wet grinding and polishing methods were used to prepare the samples for EBSD and X-ray analysis. A PANalytical Empyrean XRD instrument (Malvern Panalytical, Malvern, United Kingdom) was used with unfiltered Co-Kα radiation at 45 kV and 40 mA current for quantitative X-ray diffraction (XRD) to measure the volume fraction of phases from a 2θ spectrum that was acquired at a step size of 0.0260 over an angular range of $40°$ to $130°$. An orientation microscopy investigation of transformed austenite and martensite was conducted by electron back-scattered diffraction (EBSD) technique, using an Oxford system attached with a Carl Zeiss AURIGA® CrossBeam® (Carl Zeiss, Oberkochen, Germany) field emission gun scanning electron microscopy (FEG SEM) workstation at an accelerating voltage of 20 kV. The EBSD patterns were collected at a binning mode of 2×2 and scanning step size of 0.15 μm using AZTEC software. Orientation imaging microscopy (OIM) analysis version 8 (EDAX, AMETEK Materials Analysis Division, NJ, USA) has been used for dislocation mapping. Nano-indentation tests were carried out in load control mode on a TI 900 Hysitron Tribolab system (Hysitron, Inc., Minneapolis, USA) at load up to 8000 μN with a Berkovich three-sided pyramidal diamond tip indenter (nominal angle of $65.3°$ and radius of 200 nm). Effect of high strain rate within a single grain of RA was studied using transmission electron microscopy (TEM). At first, using the SEM we recognized the RA in deformed samples. Then the microstructure of the deformed RA was observed using a TEM equipped with a field emission gun (Philips CM 200, FEI Company, Oregon, USA) after preparing the specimens by using a dual beam FIB

(FEI xT Nova Nanolab 200, FEI Company, Hillsboro, OR, USA), and the thickness of the specimens was estimated to be ~100 nm.

3. Results and Discussion

3.1. Base Material

The initial structure of the steel samples contains martensite and ~45% RA, which was observed in the EBSD image in Figure 1a and XRD pattern in Figure 1c. The EBSD phase diagram in Figure 1a revealed the presence of RA (blue areas) with lath and plate-shaped martensite (red areas). The dislocation density, Figure 1b, close to some of the austenite-martensite grain boundaries and within the smaller grains shows a shift in the distribution towards higher dislocation, suggesting that much energy is stored in those regions. In contrast, bigger RA grains have negligible dislocation density.

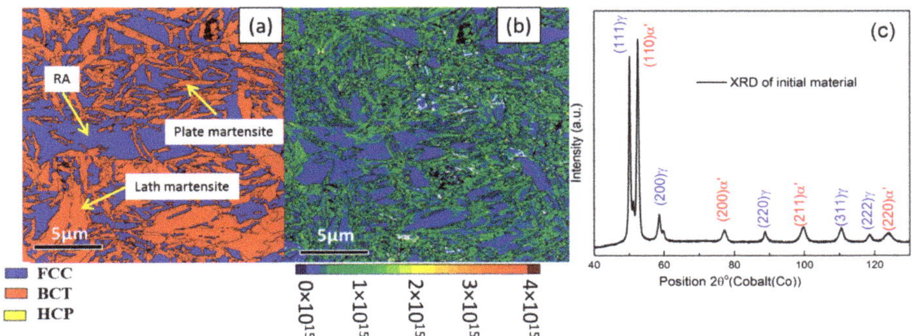

Figure 1. EBSD micrograph and XRD spectrum of the material before compression test. (**a**) Phase map, here red is martensite and blue is RA; (**b**) dislocation mapping. Black lines are grain boundaries with a misorientation angle of more than 15°; (**c**) XRD showing RA and martensite peaks.

3.2. Influence of Strain Rates on Phase Transformation

After the compression test, X-ray analysis was carried out to study the stability of phases under different compression stain rates. Figure 2a shows the XRD spectrum of different samples after compression under different strain rate and Figure 2b shows the effect of strain rates on the volume fractions of the phases. X-ray analysis was carried out to study the volume percentage of the RA of the specimen under compression deformation at different strain rates. The volume fraction can be calculated from the XRD spectrum according to the ASTM-E975–13 standard [19], where, each phase's volume fraction can be calculated based on the following Equation (1):

$$I_i^{hkl} = \frac{K R_i^{hkl} V_i}{2\mu} \tag{1}$$

where,

$$K = \left(\frac{I_0 e^4}{m^2 c^4} \right) \times \left(\frac{\lambda_A^{\,3}}{32\pi r} \right),$$

$$R_{hkl} = \left(\frac{1}{v^2} \right) \left[|F_{hkl}|^2 p \left(\frac{1 + \cos^2 2\theta}{\sin^2 \theta \cos \theta} \right) \right] \left(e^{-2M} \right).$$

$$F_{hkl} = \sum_{i=1}^{n} \sum_{j=1}^{m} k_i g_j [(f_{ij}^{\,0} + \Delta f_{ij}')^2 + \Delta f_{ij}''^{\,2}]^{1/2} \times \exp(i\{2\pi(H_{x_i} + K_{y_i} + L_{z_i}) + \arctan[\Delta f_{ij}'' / (f_{ij}^{\,0} + \Delta f_{ij}')]\}) T_{ij}^{\,iso}$$

$$M = B(\sin^2 \theta)/\lambda^2, \ B_{Fe} = 0.35 \pm \text{Å}^2.$$

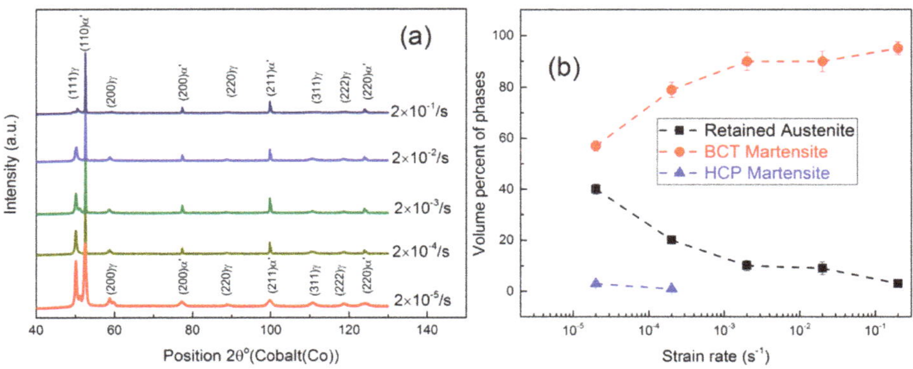

Figure 2. (**a**) X-ray diffraction of high carbon steel compressed at different strain rates at room temperature show the martensitic phase transformation of RA; (**b**) Deformation-induced samples show decreasing RA fraction and increasing martensite with increasing stress.

In this equation, I_i^{hkl} = Integrated intensity per angular diffraction peak (hkl) in the i-phase; I_0 = Intensity of the incident beam; μ = Linear absorption coefficient for the steel; e, m = Charge and mass of the electron; r = Radius of the diffractometer; c = Velocity of light; λ = Wavelength of incident radiation; v = Volume of the unit cell; F_{hkl} = Structure factor which depends on the atomic co-ordinates (x_i, y_i, z_i), f_{ij}^{0} the atomic scattering factors and anomalous dispersion corrections $(\Delta f_{ij}', \Delta f_{ij}'')$, the isotropic Debye-Waller factors T_{ij}^{iso} and the influence of occupations k_i and replacements g_j; p = Multiplicity factor of the (hkl) reflection; θ = Bragg angle; e^{-2M} = Debye-Waller or temperature factor which is a function of θ; V_i = Volume fraction of i-plane. The constant K is composed of various physical properties of the material. The terms in the R factor involve the unit cell volume, structure factor, crystallographic multiplicity factor, Lorentz polarization factor and the temperature factor.

Hence, for this high carbon steel containing RA (γ), body-centered tetragonal-martensite (α) and hexagonal close-packed-martensite (ε), the total volume of the phases can be written as:

$$V_\gamma + V_\alpha + V_\varepsilon = 1 \tag{2}$$

Based on Equation (1) individual volume fractions of each phase can be calculated using equation the following equation,

$$I_\gamma = \frac{KR_\gamma V_\gamma}{2\mu}, I_\alpha = \frac{KR_\alpha V_\alpha}{2\mu}, I_\varepsilon = \frac{KR_\varepsilon V_\varepsilon}{2\mu} \tag{3}$$

In a specific X-ray diffraction plane $\left(\frac{2\mu}{K}\right)$ is a constant, therefore,

$$V_i = \frac{\frac{1}{n}\sum_{j=1}^{n}\frac{I_i^j}{R_i^j}}{\frac{1}{n}\sum_{j=1}^{n}\frac{I_\gamma^j}{R_\gamma^j} + \frac{1}{n}\sum_{j=1}^{n}\frac{I_\alpha^j}{R_\alpha^j} + \frac{1}{n}\sum_{j=1}^{n}\frac{I_\varepsilon^j}{R_\varepsilon^j}} \tag{4}$$

where $i = \gamma, \alpha, \varepsilon$ and n = number of peaks examined by X-ray diffraction.

From the X-ray spectrum (Figure 2a), ~40% RA was measured after loading at 2×10^{-5}/s strain rate which indicates the RA did not transform significantly at this stage. However, a very small peak of ε martensite was identified at the very low strain, 2×10^{-5}/s. At the highest strain rate of 2×10^{-1}/s, the content of RA decreased to below 7%. When the RA achieves sufficient energy from the induced compressive strain, martensitic transformation takes place [20]. Because of the transformation phenomenon, the amount of RA decreases as the applied strain rate is increased. In these spectrums, the α'-martensitic transformation dominates. When the imposed strain rate was 2×10^{-3}/s, the volume percent of martensite attained the value of ~90%. At further increase in the strain rate at 2×10^{-2}/s no significant increase in martensite percentage was observed. However, when the imposed strain rate was higher than 2×10^{-2}/s, which is 2×10^{-1}/s the volume percent of retained austenite increased by ~5% and attained the value of ~95%. At this stage, RA was encapsulated in the newly formed martensite plates and the dislocation defects increases which restricted further transformation.

3.3. Influence of Strain Rates on Microstructure

The microstructure and dislocation density of martensite in different specimens that had undergone different strain rates were evaluated. The results are shown in Figure 3. Local variations in the lattice orientation reflect lattice curvature that can be associated with residual strain and geometrically necessary dislocations (GNDs) [21,22]. Different levels of dislocations were observed within the microstructure, which is indicative of the heterogeneity of strain developed through the microstructure at different strain rates of compressive loading. The heterogeneity in the microstructure also shows the phase transformation phenomena. When the RA gets sufficient energy from the compressive stress, the phase transformation takes place. This phenomenon occurs simultaneously with the increasing compressive strain rates. The highly deformed areas following compression are close to the grain boundaries, the locations for the nucleation of new martensite. This study also reveals that a small amount of ε martensite formed at the low strain rate deformation. At higher strain rate only α' martensite formation was observed. Later, when strain rate reached and exceeded 2×10^{-3}/s, the formation of the new martensite slowed down. These observations are made based on phase mapping (Figure 3(a1,b1,c1,d1,e1,f1)), which are also supported by the XRD data in Figure 2.

It is well-known from the literature that compression deformation and phase transformation both generate high dislocation density in steel. At the base material, in Figure 3(a2), the dislocation density was minimal compared to the compressed samples. As the RA phase has less dislocation density and ~45% RA was present in the microstructure (Figure 3(a1)), the overall dislocation density demonstrates less value. At the 2×10^{-5}/s strain rate (Figure 3(b2)), the overall dislocation density increased due to dislocation of newly transformed martensite. Dislocation density in the parent martensite also increased at this stage due to the increase in misorientation angles within the grains by deformation. It is noteworthy that dislocation density in the microstructure was a little bit less in 2×10^{-4}/s (Figure 3(c2)) compared to 2×10^{-5}/s (Figure 3(b2)) strain rate induced sample. This happened because to compress the sample at a low strain rate (2×10^{-5}/s) needs more time to attain the desired compressive load compared to the slightly high strain rate (2×10^{-4}/s). This extra time for deformation creates more dislocation in the microstructure. However, when the strain is very high, which is 2×10^{-1}/s, most of the RA grains transformed to martensite (Figure 3(f1)) and to accommodate the total strain the dislocation density increased further (Figure 3(f2)).

Deformation by compressive loading creates an accumulation of the dislocation density within the microstructure. When we apply the load in high strain rate, the microstructure does not have time to relax by reason of the generation of dislocation. The formation of dislocation is higher than the speed of dislocation movement which results in a much higher concentration of dislocation and more pile-ups, forms low-angle subgrain boundaries. The dislocation density increases continuously within the sub-grain boundary due to the increase in misorientation angle by high strain rate and eventually, this misorientation turns out to be high enough to create new grain boundaries. The amount of such boundaries gradually increases with the increased strain rate. The dislocation density drops with the formation of high angle boundaries, because, new boundaries are composed of dense distributions of dislocations and consumed all the dislocations, hence, the surrounding regions demonstrate a drop in dislocation content. This is the reason behind the slightly decreased dislocation density at 2×10^{-4}/s strain rate (Figure 3(c2)) compared to the 2×10^{-5}/s strain rate (Figure 3(b2)). After this strain rate, the dislocation density increases again. This phenomenon concludes in phase transformation and grain boundary refinement, as confirmed by the microstructural evolution via high-resolution EBSD analysis. If the steel structure is refined by phase transformation and increased dislocation density due to the mechanical deformation, the hardness of the structure increases. The origin of this high hardness is the reduced grain size and grain boundary strengthening results from the hindrance of grain boundaries to dislocation motion.

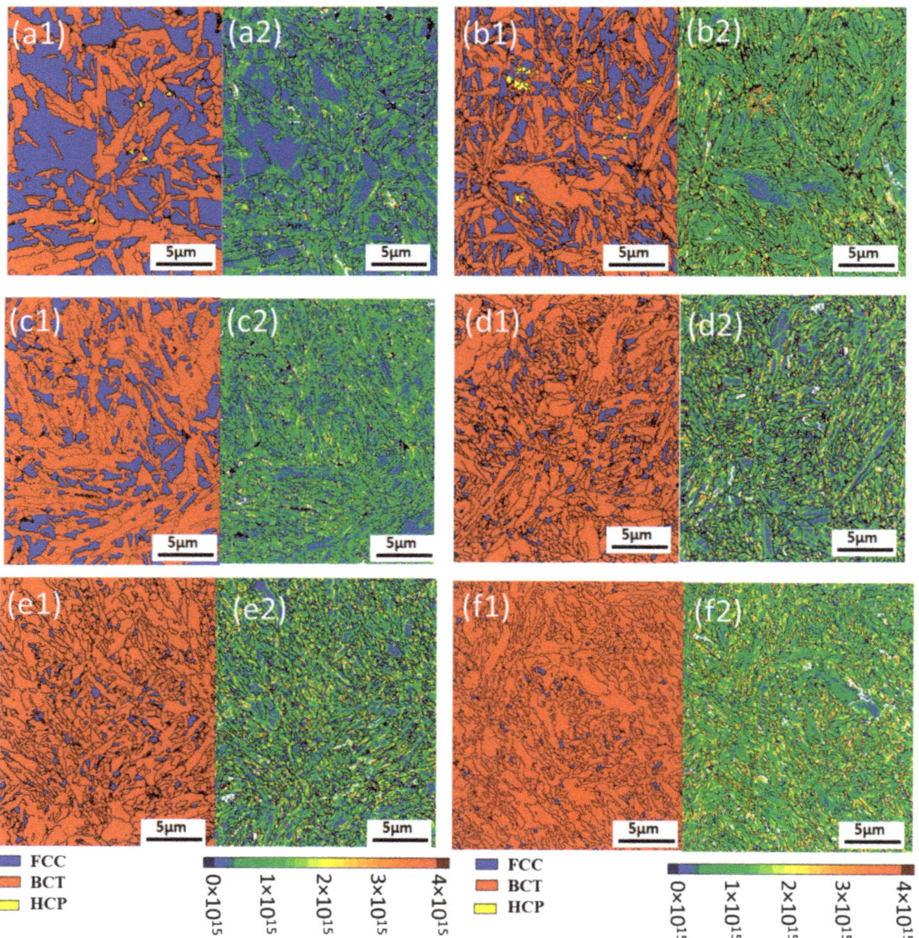

Figure 3. EBSD micrograph of the material (**a1**) phase map & (**a2**) dislocation density map of base material; (**b1**) phase map & (**b2**) dislocation density of 2×10^{-5}/s; (**c1**) phase map & (**c2**) dislocation density of 2×10^{-4}/s; (**d1**) phase map & (**d2**) dislocation density of 2×10^{-3}/s; (**e1**) phase map & (**e2**) dislocation density of 2×10^{-2}/s; (**f1**) phase map & (**f2**) dislocation density of 2×10^{-1}/s strain rate induced samples.

Apart from phase transformation and grain refinement, the deformation mechanism of steel has also involved the interactions between deformation twins and dislocations. Formation of nanotwins due to the deformation strengthens the structure and restricts the transformation of RA. It is evident from the literature that the increase in strain rate accelerates the formation of deformation twins in order to accommodate the total strain developed. Figure 4 demonstrates the TEM images where at a high strain rate of 2×10^{-2}/s dislocation interaction and deformation twins were observed simultaneously in the RA grain as shown in Figure 4c. In contrast, at low strain rate, dislocation tangles and dislocation cells are clearly visible (Figure 4a). Figure 4b,c also demonstrates the morphology of the deformation twin bundles in steel at different strain rates. As the strain rate increases, the thickness of the twins decreases. In this study, the strain hardening behavior of austenite grains is governed by two processes: dislocation-dislocation interaction hardening, and dislocation twin boundary interaction hardening which is in line with the previous literature [22].

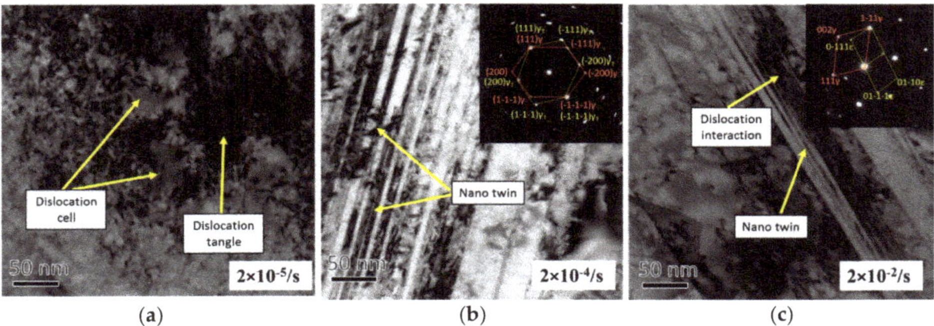

Figure 4. Presence of dislocation cell, dislocation tangles, nanotwins in deformed austenite at: (**a**) 2×10^{-5}/s, (**b**) 2×10^{-4}/s and (**c**) 2×10^{-2}/s strain rates. The SAED pattern of the nanotwins is showing orientation relationship with the parent RA grains.

In the current study, the compressive load and specimen size were the same for each sample; we deformed the sample at different strain rates: 2×10^{-5}/s, 2×10^{-4}/s, 2×10^{-3}/s, 2×10^{-2}/s and 2×10^{-1}/s. At the low strain rate, more time is needed to attain the desired compressive stress. Therefore, the twin formation process within the grains can get more time to form. In contrast, at a high strain rate, twin cannot have sufficient time to grow due to the speedy deformation. As a result, the width of the nanotwins was more at the low strain rate compared to the high strain rate (Figure 4b,c). However, the density of nanotwins at a high strain rate of 2×10^{-2}/s was found to be much higher than the nanotwins at a low strain rate of 2×10^{-4}/s. In order to accommodate the total strain, the formation of twins increases at higher strain but cannot grow to become thicker due to less time for deformation. The selected area electron diffraction (SAED) patterns of the nanotwins are showing orientation relationship in both the deformation-induced samples.

3.4. Influence of Strain Rates on Mechanical Property

It is well known that sample hardness increases with the martensitic transformation of RA and grain refinement. Martensite restricts the dislocation movements and hence deformation. The hardness of metallic materials also rises with dislocation density and nano twin formation by deformation. In this current study, plastic deformation and phase transformation both generates dislocation defects.

Figure 5 shows the results of samples in nanohardness vs. compressive stress tests. Nanohardness increases gradually with the increased strain rate and decreased grain size up to 2×10^{-3}/s. As the strain rate researches certain values between 2×10^{-3}/s and 2×10^{-2}/s, decreasing trend of the grain sizes is ceased by the saturation in phase transformation which leads to a steady-state condition of dislocation density. However, these two samples have a little difference in nanohardness. EBSD phase diagrams show a slight variation in the size of RA grains; the sample under compressive stress at 2×10^{-3}/s strain rate has slightly bigger RA grains compared to the 2×10^{-2}/s strain rate induced sample. The overall distribution of the grains also shows less variation and comparatively less grain size for 2×10^{-2}/s strain rate induced sample. The decrease in the RA grain size and overall grain size is the reason behind more hardness in the 2×10^{-2}/s strain rate induced sample. When the strain rate increased and reached 2×10^{-1}/s, a sudden increase in hardness occurred, and exceeded 10 GPa (Figure 5a,b). This correlated closely with the proportion of martensite in the structure (Figure 2), the transformation of RA to martensite (as indicated in Figure 5b), the increase in dislocation defects (Figure 3) and grain refinement (Figure 5a). Before compression deformation, the nanohardness was measured at 7.5 MPa. After compression deformation at 2×10^{-1}/s strain rate, the nanohardness increased and attained a value of 10.4 GPa. As such, the hardness of the steel was some ~35–40% higher than the hardness of the original steel sample.

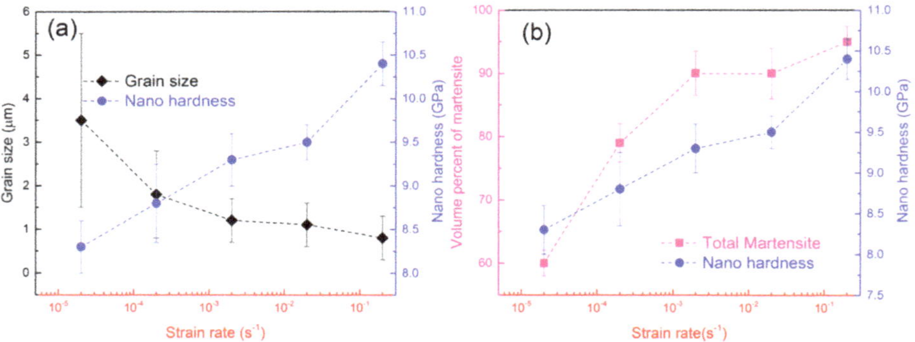

Figure 5. (**a**) Grain size profile and nanohardness under compression stress at various strain rates; (**b**) The volume percent of martensite and nanohardness profile under compressive stress at different strain rates.

4. Conclusions

This work explored the mechanical stability of RA in high carbon steel samples under compressive stress at different strain rates, using a combination of EBSD, XRD, TEM and nano-indentation tests. XRD and EBSD patterns demonstrate the volume of RA at increased compressive strain rate. The volume percentage of martensite increased as the strain rate increased, hence efficiently increases the strength of the structure. EBSD showed a significant reduction in the grain size. Nanohardness tests showed a positive correlation between strain rates and the hardness of the material, which can be attributed to strain hardening and phase transformation.

We have established and comprehensively described how higher strain rate induced plastic deformation produced refine grains of metals and make them very strong. When heavily deformed through compressive loading, metal specimens generated an increasing density of defects or dislocations that effectively strengthens the metal against further deformation. The extraordinary hardness is achieved by simultaneously enabling transformation-induced and dislocation density-induced structural strengthening mechanisms. Another nanoscale mechanism involves strengthening by nano twinning. Such research is essential for controlling the microstructures of high carbon steels under extreme operation condition and, so, for opening up new industrial applications for these relatively low alloyed low-cost steel.

Author Contributions: Conceptualization, R.H.; Methodology, R.H.; Formal Analysis, R.H.; Investigation, R.H.; Writing—Original Draft Preparation, R.H.; Writing—Review & Editing, R.H., F.P. and V.S.; Supervision, F.P. and V.S.; Project Administration, V.S.; Funding Acquisition, V.S.

Funding: This research was funded by Australian Research Council's Industrial Transformation Research Hub funding scheme, project IH130200025.

Acknowledgments: This research was supported under Australian Research Council's Industrial Transformation Research Hub funding scheme (project IH130200025). We gratefully acknowledge the technical support provided by the Analytical Centre in the UNSW Australia.

Conflicts of Interest: The authors declare no conflict of interest.

References

1. Hossain, R.; Pahlevani, F.; Quadir, M.; Sahajwalla, V. Stability of retained austenite in high carbon steel under compressive stress: An investigation from macro to nano scale. *Sci. Rep.* **2016**, *6*, 34958. [CrossRef] [PubMed]
2. Hossain, R.; Pahlevani, F.; Sahajwalla, V. Solid State Phase Transformation Mechanism in High Carbon Steel Under Compressive Load and with Varying Cr Percent. In Proceedings of the TMS Annual Meeting & Exhibition, Phoenix, AZ, USA, 11–15 March 2018; pp. 797–802.
3. Kim, J.H.; Ko, K.H.; Noh, S.D.; Kim, G.G.; Kim, S.J. The effect of boron on the abrasive wear behavior of austenitic Fe-based hardfacing alloys. *Wear* **2009**, *267*, 1415–1419. [CrossRef]
4. Hossain, R.; Pahlevani, F.; Witteveen, E.; Banerjee, A.; Joe, B.; Prusty, B.G.; Dippenaar, R.; Sahajwalla, V. Hybrid structure of white layer in high carbon steel—Formation mechanism and its properties. *Sci. Rep.* **2017**, *7*, 13288. [CrossRef] [PubMed]
5. Celada-Casero, C.; Kooiker, H.; Groen, M.; Post, J.; San-Martin, D. In-situ investigation of strain-induced martensitic transformation kinetics in an austenitic stainless steel by inductive measurements. *Metals* **2017**, *7*, 271. [CrossRef]
6. Hossain, R.; Pahlevani, F.; Sahajwalla, V. Effect of small addition of Cr on stability of retained austenite in high carbon steel. *Mater. Charact.* **2017**, *125*, 114–122. [CrossRef]
7. Jeon, J.B.; Chang, Y.W. Effect of nitrogen on deformation-induced martensitic transformation in an austenitic 301 stainless steels. *Metals* **2017**, *7*, 503. [CrossRef]
8. Curtze, S.; Kuokkala, V.-T. Dependence of tensile deformation behavior of TWIP steels on stacking fault energy, temperature and strain rate. *Acta Mater.* **2010**, *58*, 5129–5141. [CrossRef]
9. He, B.; Huang, M.; Liang, Z.; Ngan, A.; Luo, H.; Shi, J.; Cao, W.; Dong, H. Nanoindentation investigation on the mechanical stability of individual austenite grains in a medium-Mn transformation-induced plasticity steel. *Scr. Mater.* **2013**, *69*, 215–218. [CrossRef]
10. Qiao, X.; Han, L.; Zhang, W.; Gu, J. Nano-indentation investigation on the mechanical stability of individual austenite in high-carbon steel. *Mater. Charact.* **2015**, *110*, 86–93. [CrossRef]
11. He, B.; Luo, H.; Huang, M. Experimental investigation on a novel medium Mn steel combining transformation-induced plasticity and twinning-induced plasticity effects. *Int. J. Plast.* **2016**, *78*, 173–186. [CrossRef]
12. Ahn, T.-H.; Oh, C.-S.; Kim, D.; Oh, K.; Bei, H.; George, E.P.; Han, H. Investigation of strain-induced martensitic transformation in metastable austenite using nanoindentation. *Scr. Mater.* **2010**, *63*, 540–543. [CrossRef]
13. Hall, E. The deformation and ageing of mild steel: III discussion of results. *Proc. Phys. Soc. Sect. B* **1951**, *64*, 747–753. [CrossRef]
14. Lu, L.; Chen, X.; Huang, X.; Lu, K. Revealing the maximum strength in nanotwinned copper. *Science* **2009**, *323*, 607–610. [CrossRef] [PubMed]
15. Cadoni, E.; Fenu, L.; Forni, D. Strain rate behaviour in tension of austenitic stainless steel used for reinforcing bars. *Constr. Build. Mater.* **2012**, *35*, 399–407. [CrossRef]
16. Kim, C. Nondestructive evaluation of strain-induced phase transformation and damage accumulation in austenitic stainless steel subjected to cyclic loading. *Metals* **2017**, *8*, 14. [CrossRef]
17. Nakkalil, R. Formation of adiabatic shear bands in eutectoid steels in high strain rate compression. *Acta Metallurgica et Materialia* **1991**, *39*, 2553–2563. [CrossRef]
18. Standard, A. *E975-03: Standard Practice for X-ray Determination of Retained Austenite in Steel with Near Random Crystallographic Orientation*; ASTM: West Conshohocken, PA, USA, 2008.
19. Böhner, A.; Niendorf, T.; Amberger, D.; Höppel, H.W.; Göken, M.; Maier, H.J. Martensitic transformation in ultrafine-grained stainless steel AISI 304L under monotonic and cyclic loading. *Metals* **2012**, *2*, 56–64. [CrossRef]

20. Li, H.; Hsu, E.; Szpunar, J.; Utsunomiya, H.; Sakai, T. Deformation mechanism and texture and microstructure evolution during high-speed rolling of AZ31B Mg sheets. *J. Mater. Sci.* **2008**, *43*, 7148–7156. [CrossRef]
21. Gouné, M.; Danoix, F.; Allain, S.; Bouaziz, O. Unambiguous carbon partitioning from martensite to austenite in Fe-C-Ni alloys during quenching and partitioning. *Scr. Mater.* **2013**, *68*, 1004–1007. [CrossRef]
22. Kou, H.; Lu, J.; Li, Y. High-strength and high-ductility nanostructured and amorphous metallic materials. *Adv. Mater.* **2014**, *26*, 5518–5524. [CrossRef] [PubMed]

Article

Effect of Deformation Temperature on Mechanical Properties and Deformation Mechanisms of Cold-Rolled Low C High Mn TRIP/TWIP Steel

Zhengyou Tang [1,*], Jianeng Huang [1,*], Hua Ding [1], Zhihui Cai [1], Dongmei Zhang [2] and Devesh Misra [3]

[1] School of Materials Science and Engineering, Northeastern University, Shenyang 110819, China; neuhd@sina.com (H.D.); neuczh@sina.com (Z.C.)

[2] Shenyang Liming Aero-Engine Corporation, AECC, Shenyang 110043, China; 15040279961@163.com

[3] Laboratory for Excellence in Advanced Steel Research, Department of Metallurgical, Materials and Biomedical Engineering, University of Texas at El Paso, El Paso, TX 79968, USA; dmisra2@utep.edu

* Correspondence: tangzy@smm.neu.edu.cn (Z.T.); Huangjnd@163.com (J.H.)

Received: 22 May 2018; Accepted: 20 June 2018; Published: 22 June 2018

Abstract: The microstructure and mechanical properties of cold-rolled Fe-18Mn-3Al-3Si-0.03C transformation induced plasticity/twinning induced plasticity (TRIP/TWIP) steel in the temperature range of 25 to 600 °C were studied. The experimental steel exhibited a good combination of ultimate tensile strength (UTS) of 905 MPa and total elongation (TEL) of 55% at room temperature. With the increase of deformation temperature from 25 to 600 °C, the stacking fault energy (SFE) of the experimental steel increased from 14.5 to 98.8 mJm^{-2}. The deformation mechanism of the experimental steel is controlled by both the strain induced martensite formation and strain induced deformation twinning at 25 °C. With the increase of deformation temperature from 25 to 600 °C, TRIP and TWIP effect were inhibited, and dislocation glide gradually became the main deformation mechanism. The UTS decreased monotonously from 905 to 325 MPa and the TEL decreased (from 55 to 36%, 25–400 °C) and then increased (from 36 to 64%, 400–600 °C). The change in mechanical properties is related to the thermal softening effect, TRIP effect, TWIP effect, DSA, and dislocation slip.

Keywords: deformation temperature; deformation mechanisms; tensile properties; TRIP/TWIP steel

1. Introduction

Low C, high Mn transformation-induced plasticity/twinning-induced plasticity (TRIP/TWIP) steels are considered to be one of the most attractive materials for automotive steels because of their excellent combination of strength (>900 MPa) and ductility (>50%) at room temperature [1–5]. The outstanding mechanical properties of TRIP/TWIP steels at room temperature are due to the remarkable work-hardening behavior resulting from the evolution of multiple microstructural processes including dislocation slip, formation of stacking fault, deformation induced martensitic transformation and deformation twinning [3–7]. The $\alpha_{bcc}/\varepsilon_{hcp}$-martensite and mechanical twins (transformed from austenite during deformation) act as planar obstacles and reduce the mean free path of dislocation glide, promoting working hardening and delaying necking, which results in large uniform elongation [5,7,8]. Recently, it was proposed that the influence of dynamic strain aging (DSA) phenomenon on the mechanical behavior of high Mn TWIP steels should not be ignored, and DSA phenomena may occur in a particular temperature range [9–12]. The loss of ductility caused by DSA has been reported for ferritic steel, low carbon steel, and dual-phase steel at the DSA temperature range [13–16]. A change of temperature can influence DSA [17]. Thus, considering the complexity of deformation mechanisms of TRIP/TWIP steel, it is necessary to understand the deformation mechanisms of TRIP/TWIP steels at different temperatures.

The mechanical properties and deformation mechanisms of TRIP/TWIP steels exhibit a strong dependence on temperature and stacking fault energy (SFE) [8,18]. It is well established that the main governing factor responsible for the deformation mechanism in TRIP/TWIP steels is the SFE [19] and the SFE increased with the increased of temperature [7,20]. According to Curtze et al. [21], when SFE below 18 mJm^{-2}, strain induced martensite transformation is the favored transformation mechanism. When the SFE increases from ~18 to 45 mJm^{-2}, deformation-induced twinning occurs [21,22]. When SFE exceeds 45 mJm^{-2}, dislocation glide controls the work hardening [18,21]. The study of SFE for specific plasticity mechanisms is important because deformation mode can affect the mechanical properties of materials or result in deleterious effects. For example, DSA in some TRIP/TWIP steels [8,11,17,23].

A few studies were devoted to TRIP/TWIP steels deformed at different temperatures [7,18,22,24]. Linderov et al. [7] investigated the deformation mechanisms of austenitic TRIP/TWIP steels at two different temperatures—ambient and 373 K (100 °C)—and observed that the temperature strongly affects the deformation-induced processes of metastable austenitic steels. However, a systematic study on the effect of deformation temperature on the deformation mechanism and mechanical properties of TRIP/TWIP steels was not carried out. Martin et al. [24] studied the deformation mechanism in a high-alloy austenitic CrMnNi austenitic TRIP/TWIP steel in the temperature range of −60 and 200 °C, and demonstrated that the change in deformation mechanism was caused by both temperature and SFE. Asghari et al. [19] investigated a TRIP/TWIP steel (Fe–0.07C–18Mn–2Si–2Al) over a wide temperature range from 25 to 1000 °C through compression tests, and showed that the flow stress was strongly dependent on the deformation temperature. Eskandari et al. [25] also conducted compressive tests on a TRIP/TWIP steel (Fe–0.11C–21Mn–2.70Si–1.60Al–0.01Nb–0.01Ti–0.10Cr) from 25 to 1000 °C to investigate the mechanical behavior of the TRIP/TWIP steel. They concluded that both the strain-induced martensite formation and mechanical twinning controlled the plastic deformation in the temperature range of 25–150 °C. In summary, a systematic study on the influence of deformation temperature (over a wide range of temperature) on the deformation mechanism and tensile properties of TRIP/TWIP steel is required at this time.

Thus, The present study is aimed at studying the effect of deformation temperature (over a temperature range of 25 to 600 °C) on the deformation mechanism and tensile properties of TRIP/TWIP steel to get a comprehensive understanding of warm/hot deformation characteristics of TRIP/TWIP steel, having a better insight into the evolution of deformation mechanism in TRIP/TWIP steel at different temperatures, which can provide reference for the best design of some processing route such as warm stamping, hot rolling, thermo-mechanical process. The experimental results have a certain practical significance in the deformation technology of TRIP/TWIP steel in different temperature ranges.

2. Materials and Methods

The nominal chemical composition of the investigated TRIP/TWIP steel was Fe-18.10Mn-3.1Al-3.2Si-0.03C (in wt%). A 50 kg ingot was cast after melting the steel in a vacuum furnace. The ingot was heated at 1200 °C for 2 h and hot forged to rods of section size 100 mm × 30 mm, followed by air cooling to room temperature. Subsequently, the rod was soaked at 1200 °C for 2 h, then hot rolled to 3 mm thickness. The hot rolled plate was reheated to 1100 °C for 30 min and cold-rolled to sheet of 1 mm thickness. Finally, the cold rolled sheet was solution-treated at 1000 °C for 10 min and water-quenched. Tensile specimens with a gage length of 25 mm and width of 6 mm were machined from the cold rolled sheet with the tensile axis parallel to the rolling direction. Tensile tests were conducted from 25 to 600 °C (25 (room temperature), 200, 300, 400, 500, and 600 °C, referred to as 25-sample, 200-sample, 300-sample, 400-sample, 500-sample, and 600-sample, respectively) using CMT5105 tensile testing machine at a constant strain rate of 1.67×10^{-3} s^{-1}. In order to heat the specimen uniformly, high-temperature deformation testing was carried out in a sealed insulation furnace, and a heat-resistant alloy fixture was used. The sample was heated to the deformation temperature at the rate of 10 °C/s and maintained at the test temperature for 5 min.

Microstructure was studied by optical microscope (OM, OLYMPUS-GSX500, OLYMPUS, Tokyo, Japan), field-emission scanning electron microscope (FE-SEM, Supra, SSX-550, Shimadzu, Tokyo, Japan) and a field-emission transmission electron microscope (FE-TEM, FEI, TECNAI G2-20, operated at 200 kV, FEI, Hillsboro, OR, USA). Samples were cut from the tensile specimens near the fracture surface after the tensile tests. Samples were mechanically polished and subsequently electropolished at room temperature in an electrolyte containing 95% CH_3COOH+5% $HClO_4$ solution to remove any strain-induced martensite that may have formed during mechanical polishing. Phase analysis was carried out by X-ray diffraction (XRD, Rigaku, D/Max2250/PC, Rigaku Corporation, Tokyo, Japan) with CuK_α radiation. Before OM and SEM observations, the samples were etched with 4% nital. For TEM studies, thin foils were first mechanically ground to ~40 µm thickness, then twin-jet polished (Struers, Tenupol-5, Struers, Copenhagen, Denmark) in a solution of 5% perchloric acid and 95% alcohol at ~20 °C. Annealing and deformation twins, stacking faults, martensite, dislocation walls, and slip bands were identified using a TECNAI G2-20 TEM.

3. Results

3.1. Mechanical Properties and Work Hardening Behavior

Figure 1 shows engineering stress-strain curves of the experimental steel at different tensile test temperatures. The statistical results of ultimate tensile strength (UTS), total elongation (TEL), and the product of ultimate tensile strength and total elongation (PSE) of the experimental steel at different deformation temperatures are summarized in Table 1. Combining Figure 1 and Table 1, it may be noted that the deformation temperature has a significant influence on the mechanical behavior of the investigated steel. The thermal softening effect is obvious with the increase of temperature. The UTS and PSE continuously decreased with the increase of temperature from 25 to 600 °C, a behavior similar to the results of Shterner et al. [3]. The TEL decreased from 55 to 36%, as the temperature increased from 25 to 200 °C. Interestingly, TEL remained almost unchanged at 36%, when the deformation temperature was in the range of 200 to 400 °C. Then, TEL increased from 36 to 64% with increase of tensile test temperature from 400 to 600 °C.

It may be noted that the periodic serration and step-like fluctuation appeared in the engineering stress-strain curves, when the deformation temperature was in the range of 200 to 400 °C (Figure 1). According to the conclusion by Lee et al. [11] and Lan et al. [26], the periodic serration and step-like fluctuation are representative of DSA and Portevin Le-Chatelier (PLC) behavior. However, DSA did not appear when the tensile tests were carried out at other temperatures. This indicates that the experimental steel is most susceptible to experience DSA in the temperature range of 200–400 °C.

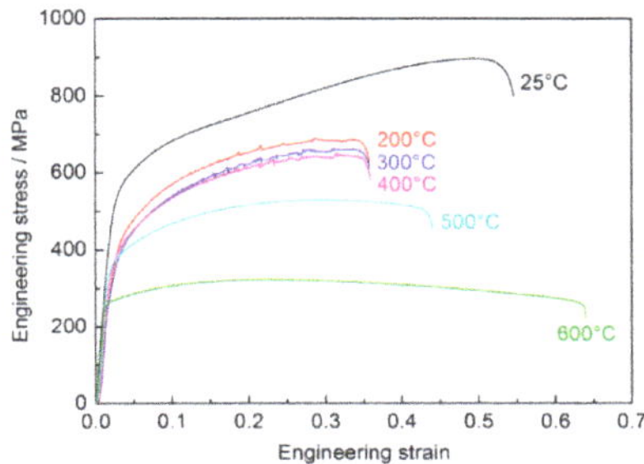

Figure 1. Engineering strain-stress curves at different tensile test temperatures of the experimental steel.

Table 1. Mechanical properties of experimental steel at different tensile test temperatures.

Temperature (°C)	UTS (MPa)	TEL (%)	PSE (GPa%)
25	905 (±7)	55 (±0.2)	49.8 (±0.5)
200	690 (±10)	36 (±0.3)	24.8 (±0.6)
300	670 (±6)	36 (±0.3)	24.1 (±0.4)
400	650 (±8)	36 (±0.2)	23.4 (±0.4)
500	530 (±6)	44 (±0.5)	23.3 (±0.5)
600	325 (±3)	64 (±0.6)	20.8 (±0.4)

Figure 2 presents the instantaneous work hardening exponent (n) and true stress as a function of true strain, where the n value was calculated by n = d lnσ/d lnε [27]. The instantaneous work hardening exponent (n) can further reveal the deformation behavior of the experimental steel during straining. Based on Figure 2, it can be seen that the variation in n value was different in samples deformed at different temperatures. The instantaneous work hardening exponent (n) curves can be divided into three stages. In summary, the n value decreased sharply in S_1 (elastic stage) for all the samples. During S_2, a rapid increase in the n value of 25-sample indicated a high strain hardening rate. A strong fluctuation in the n value was observed for 200-sample, 300-sample and 400-sample, and a gradual increase in the n value for 500-sample and 600-sample. Finally, the n value of 25-sample, 200-sample, 300-sample, and 400-sample decreased rapidly in S_3, while, for 500-sample and 600-sample, the decrease was slow.

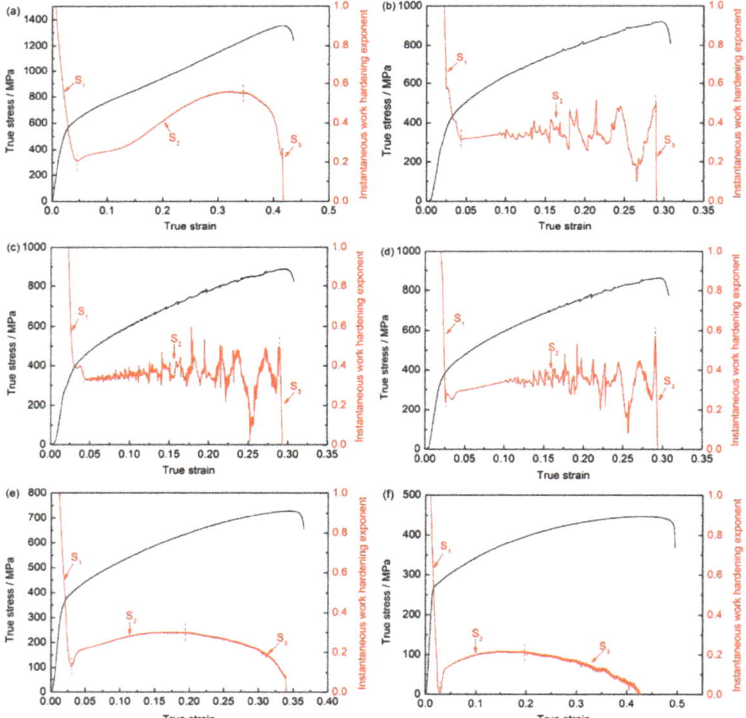

Figure 2. Instantaneous work hardening exponent (n) and true stress versus true strain of the experimental steel after deformation at different temperatures of (**a**) 25 °C; (**b**) 200 °C; (**c**) 300 °C; (**d**) 400 °C; (**e**) 500 °C; and (**f**) 600 °C. (S_1~S_3 mean stage 1–3).

3.2. Microstructure

Figure 3 shows the X-ray diffraction patterns of the investigated samples before and after tensile deformation at different temperatures. With reference to the work of Egea et al. [28], the percentage of volume fraction of each phase for each temperature configuration has been calculated and summarized in Table 2. It can be seen from Figure 3 and Table 2 that the microstructure of experimental sample before the tensile test was mainly composed of γ-fcc phase (austenite, 66.4%), α-bcc phase (ferrite, 25.8%) and some ε-hcp phase (ε-martensite, 7.8%). After deformation at 25 °C, the peaks of 111_γ, 200_γ, 220_γ, and 311_γ for γ-fcc phase became weak (the volume fraction of austenite decreased to 51.6%), but the diffraction peaks of 110_α, 200_α, and 211_α for α-bcc phase were enhanced (the volume fraction of α'-martensite increased to 15.8%), which indicates that significant TRIP effect took place during tensile deformation at 25 °C. With the increase of deformation temperature from 300 to 600 °C, the peaks of 110_ε and 002_ε for ε-hcp phase disappeared, and the amount of austenite increased to ~73.3% (see Table 2). This may be caused by the reversion from ε-martensite to austenite during tensile deformation.

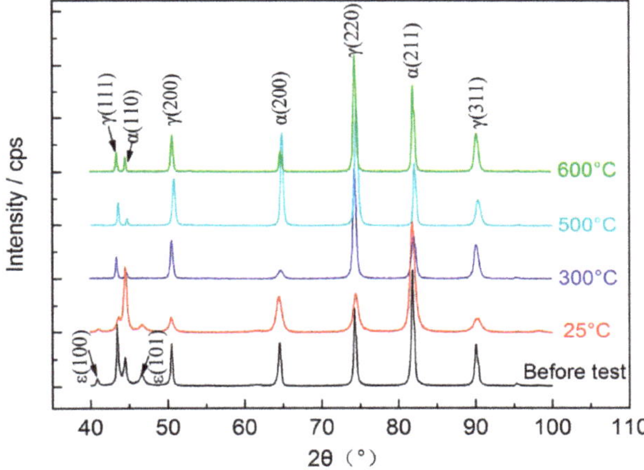

Figure 3. XRD patterns of the investigated steel before and after tensile deformation at different temperatures. ((Before test, ε = 0), (25 °C, ε = 55%), (300 °C, ε = 36%), (500 °C, ε = 44%), (600 °C, ε = 64%)).

Table 2. Volume fraction of constituent phases estimated by X-ray diffraction (vol%).

Sample	γ	α_F	ε	α'
Undeformed sample	66.4	25.8	7.8	0
25-sample	51.6	25.7	6.9	15.8
300-sample	73.3	26.7	0	0
500-sample	74.1	25.9	0	0
600-sample	73.2	26.8	0	0

The OM and SEM micrograph of the experimental steel before deformation are presented in Figure 4. As shown in Figure 4, the annealing twins were present in majority of austenite grains of the sample before the tensile test, and most of the annealing twins were present across the entire austenite grains. After deformation at room temperature, thin lamellar deformation twins (or α'-martensite) nucleated in the austenite matrix and the austenite grains were divided and refined (see Figure 5a). It can be seen from Figure 5b–d that the volume fraction of mechanical twins and α-martensite were remarkably reduced when the deformation temperature was increased to 300–600 °C. This means that the TRIP effect and TWIP effect were inhibited with the increase of temperature.

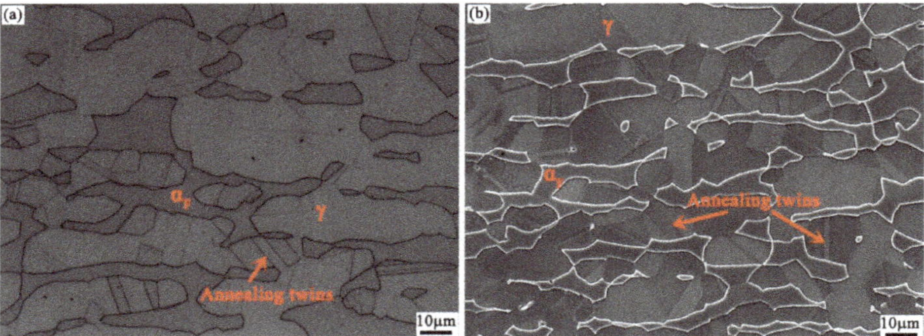

Figure 4. OM (**a**) and SEM (**b**) micrographs of the experimental steel before deformation at 25 °C ($\varepsilon = 0$).

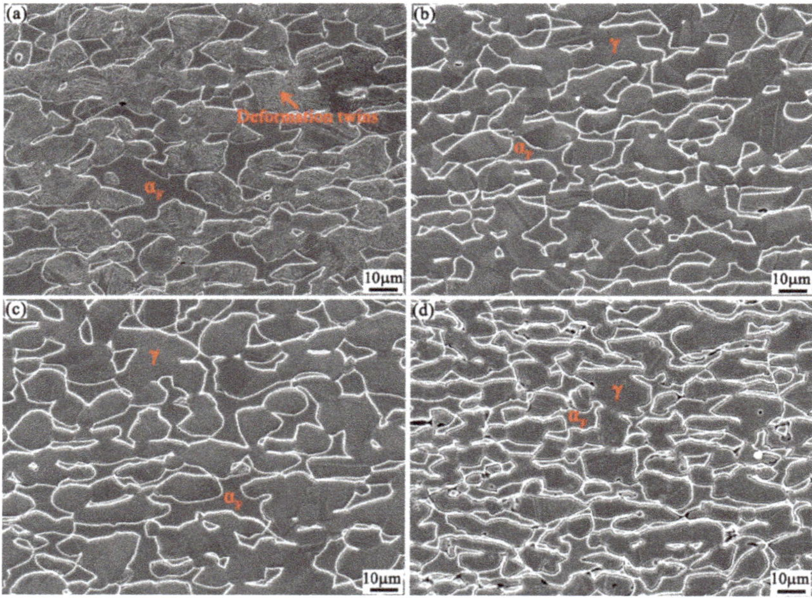

Figure 5. SEM micrographs of the experimental steel after deformation at different temperatures of (**a**) 25 °C, $\varepsilon = 55\%$; (**b**) 300 °C, $\varepsilon = 36\%$; (**c**) 500 °C, $\varepsilon = 44\%$; and (**d**) 600 °C, $\varepsilon = 64\%$.

TEM studies were conducted to reveal stacking faults, deformation twins, α'-martensite and the dislocation substructure development at different temperatures of deformation. Before deformation, the annealing twins (Figure 6a) and abundant stacking faults (Figure 6b) were present in the austenite matrix of experimental steel. The abundant stacking faults can provide favorable conditions for the subsequent strain-induced nucleation of martensite and deformation twins [4]. On deformation at room temperature (25 °C), deformation twins were observed in the austenite grain (Figure 7a). Meanwhile, α'-martensite (confirmed by electron diffraction pattern analysis) transformed from austenite during deformation (shown in Figure 7b), which indicated that the deformation-induced twins (TWIP effect) coexisted with deformation-induced α'-martensite (TRIP effect) and was the predominant deformation mechanism during tensile deformation at room temperature. When the tensile deformation temperature was increased to 300 °C, some dislocation activity was observed with a number of dislocations and dislocation walls (Figure 7c,d). This may be because the temperature of the steel was increased significantly, which led to the increase of SFE, and TRIP and TWIP effect were inhibited, but the dislocation slip was favored [7,29]. When the deformation temperature was 600 °C, dislocation cells and slip bands were observed (Figure 7e,f). Thus, dislocation slip was the main deformation mechanism in the high deformation temperature region.

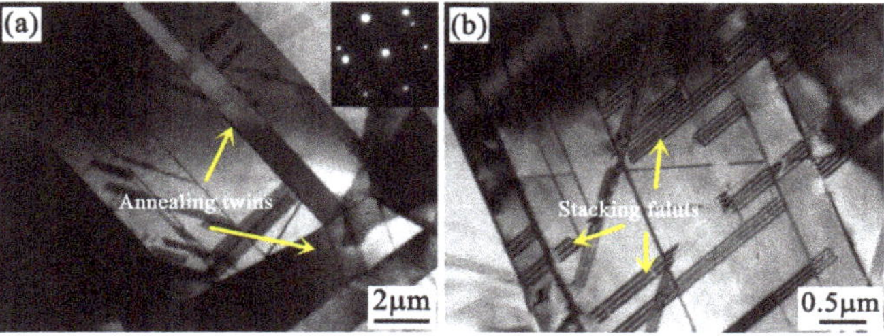

Figure 6. TEM micrographs of the experimental steel before deformation at 25 °C ($\varepsilon = 0$).

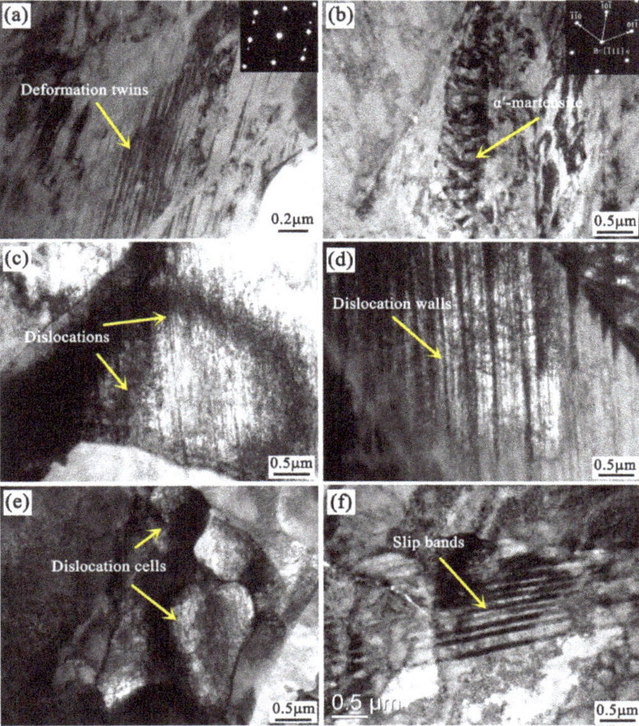

Figure 7. TEM micrographs of the experimental steel after deformation at different temperatures of (**a,b**) 25 °C, $\varepsilon = 55\%$; (**c,d**) 300 °C, $\varepsilon = 36\%$; and (**e,f**) 600 °C, $\varepsilon = 64\%$.

4. Discussion

4.1. Effect of Deformation Temperature on the Deformation Mechanism

It is known that the deformation temperature has a significant influence on microstructural development during deformation of experimental steel. The temperature change changes SFE, and the deformation behavior of high Mn steels is strongly related to SFE [9,30]. It is well known that the SFE increases with temperature [7,20]. For tests at different temperatures, the temperature rise (ΔT) is enough to affect the SFE. According to the thermodynamic models [31], the value of SFE can be calculated by the equation proposed by Saeed-Akbari et al. [20]. The change in SFE caused by ΔT is shown in Figure 8. It can be seen that the SFE increased with the increase of deformation temperature. The SFE value of the investigated steel at room temperature was 14.5 mJm^{-2}. With the increase of temperature from 200 to 600 °C, the SFE approached 40.0, 54.7, 69.4, 84.1, and 98.8 mJm^{-2}, respectively. According to Allain et al. [6] martensitic transformation occurs below 18 mJm^{-2} and twinning between 12 and 35 mJm^{-2} in the Fe–Mn–C TWIP steel. Curtze et al. [21] reported that twinning occurred at SFE: of ~18 mJm$^{-2} \leq$ SFE $\leq$ 45 mJm^{-2}. When SFE > 45 mJm^{-2}, plasticity and strain hardening were controlled solely by the glide of dislocations. Thus, the strain induced martensitic transformation and deformation twinning coexisted at the room temperature deformation because the SFE (14.5 mJm^{-2}) is relatively low (see Figures 5a and 7a,b). Then, the SFE increased with increase in temperature, TRIP effect and TWIP effect were suppressed (see Figure 5b–d). When the temperature was greater than 300 °C, twinning was replaced by dislocation slip because the SFE was greater than 45 mJm^{-2} (see Figure 7c–f).

In addition, the serrated flow (characteristic of DSA) occurred between 200 and 400 °C (see Figure 1) and implies that DSA effect has a significant effect on the mechanical properties of experimental steel in the temperature range of 200–400 °C, which is similar to the results of Bayramin et al. [16]. The SFE was in the range of 40.0 to 69.4 mJm^{-2} with the increase of temperature from 200–400 °C. Dislocation slip replaced TRIP/TWIP effect in this temperature range. The moving dislocations were pinned by solute atoms resulting in serrated flow at 200–400 °C [32]. If temperature is too high or too low, solute interstitial atoms (C-atom) may not be able to rearrange themselves with mobile dislocations and hinder them, therefore, the DSA did not take place [15,33].

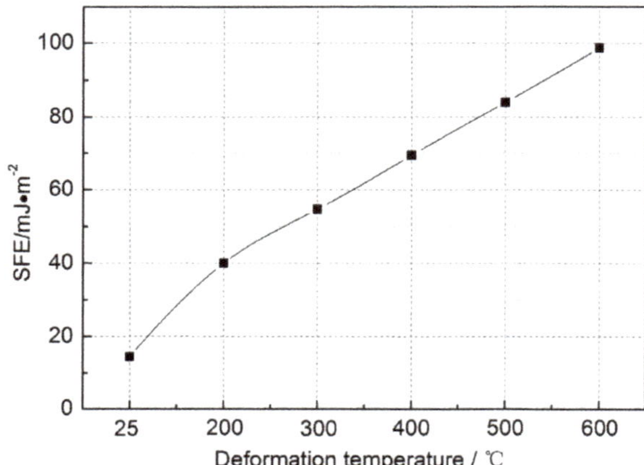

Figure 8. Estimation of SFE with the different deformation temperatures.

4.2. Tensile Properties Difference at Different Temperatures

It may be noted from Figure 1 and Table 1 that the experimental steel presents an exceptional mechanical properties with a PSE of 49.3 GPa at room temperature, and the mechanical properties deteriorated with the increase of deformation temperature. The mechanical properties are directly related to the work hardening behavior depending on the mechanisms. 25-sample exhibited the highest UTS (905 MPa) and a good plasticity with the TEL of 55%, and a rapid increase in the n value in S_2 (see Figure 2a), because of strain induced martensitic transformation (TRIP effect) and strain induced deformation twinning (TWIP effect) (see Figure 3, Figure 5a, and Figure 7a,b). The formation of martensite and deformation twins act as planar obstacles to dislocation glide during deformation and enhanced the n values (high work hardening rate), delayed necking, which improves the plasticity of the experimental steel and results in highest UTS [2,3,18,34]. It should be also noted that the increase of n value of 25-sample exceeded other samples (see Figure 2). It can be deduced that the TRIP effect and TWIP effect had the most beneficial effect on work hardening among all the deformation modes. With increase of deformation temperature to 200, 300, and 400 °C, the UTS of 200-sample, 300-sample, and 400-sample decreased to 690, 670, and 650 MPa, respectively. Meanwhile, the TEL was only ~36%. As shown in Figure 2b–d, the n value for the three samples exhibited an intense fluctuation in S_2 resulting from DSA. The decrease of UTS was due to thermal softening effect. TRIP effect and TWIP effect were suppressed and gradually replaced by dislocation slip with the increase in temperature. In this temperature range, high dislocation density and highly dense dislocation walls (see Figure 7c,d), were the main contributors in enhancing work hardening behavior. Their strengthening effect is weaker than TRIP effect and TWIP effect, hence, UTS was decreased. As regards, low plasticity (TEL of 36%), it is because the TRIP effect and TWIP effect were suppressed. In addition, the role of DSA phenomenon should be taken into account. The DSA took place in this temperature range, and an inhomogeneous plastic deformation because of successive locking of the mobile dislocations and strain localization took place, which prevented high TEL to be obtained [15]. When the deformation temperature was increased to 500–600 °C, high-temperature softening occurred and the UTS of samples was further decreased with the increase of elongation.

5. Conclusions

In the present study, the microstructure and mechanical properties of cold-rolled Fe-18Mn-3Al-3Si-0.03C TRIP/TWIP steel in the temperatures range of 25 to 600 °C was studied. The main conclusions are as follows:

1. Fe-18.1Mn-3.1Al-3.2Si-0.03C TRIP/TWIP steel exhibited excellent mechanical properties at room temperature. The UTS, TEL and PSE were 905 MPa, 55% and 49.3 GPa, respectively. The dominant plasticity mechanism at room temperature was strain induced martensite deformation and deformation twinning.

2. With the increase of deformation temperature from 25 (room temperature) to 600 °C, the SFE of the experimental steel was in the range of 14.5 to 98.8 mJm^{-2}. The deformation mechanism of the experimental steel is controlled by both the strain induced martensite formation and strain induced deformation twinning at 25 °C. With the increase of deformation temperature from 25 to 600 °C, TRIP and TWIP effects were inhibited, and dislocation glide gradually became the main deformation mechanism.

3. The UTS decreased monotonously from 905 to 325 MPa and the TEL decreased (from 55 to 36%, 25–400 °C) and then increased (from 36 to 64%, 400–600 °C). The change in mechanical properties was related to the thermal softening effect, TRIP effect, TWIP effect, DSA, and dislocation slip.

Author Contributions: Z.T. and J.H. designed most of the experiments, performed most experiments, analyzed the results, and wrote this manuscript. Z.C., H.D., D.Z., and D.M. helped analyze the experiment data and gave some constructive suggestions about how to write this manuscript.

Funding: This research received no external funding.

Acknowledgments: The research was supported by the National Natural Science Foundation of China (grant Nos. 51574077 and 51501035) and the Fundamental Research Funds for the Central Universities (N120502005 and N170204017). R.D.K. Misra acknowledges continued collaboration with Northeastern University as an Honorary Professor in providing guidance to students in research.

Conflicts of Interest: The authors declare no conflict of interest.

References

1. Ding, H.; Tang, Z.Y.; Li, W.; Wang, M.; Song, D. Microstructures and mechanical properties of Fe-Mn-(Al, Si) TRIP/TWIP steels. *J. Iron Steel Res.* **2006**, *13*, 66–70. [CrossRef]
2. Ding, H.; Ding, H.; Song, D.; Tang, Z.Y.; Yang, P. Strain hardening behavior of a TRIP/TWIP steel with 18.8% Mn. *Mater. Sci. Eng. A* **2011**, *528*, 868–873. [CrossRef]
3. Shterner, V.; Timokhina, I.B.; Beladi, H. On the work-hardening behaviour of a high manganese TWIP steel at different deformation temperatures. *Mater. Sci. Eng. A* **2016**, *669*, 437–446. [CrossRef]
4. Tang, Z.Y.; Misra, R.D.K.; Ma, M.; Zan, N.; Wu, Z.Q.; Ding, H. Deformation twinning and martensitic transformation and dynamic mechanical properties in Fe-0.07C-23Mn-3.1Si-2.8Al TRIP/TWIP steel. *Mater. Sci. Eng. A* **2015**, *624*, 186–192. [CrossRef]
5. Dufay, A.; Chateau, J.P.; Allain, S.; Migot, S.; Bouaziz, O. Influence of addition elements on the stacking-fault energy and mechanical properties of an austenitic Fe-Mn-C steel. *Mater. Sci. Eng. A* **2008**, *483–484*, 184–187.
6. Allain, S.; Chaterau, J.P.; Bouaziz, O.; Migot, S.; Guelton, N. Correlations between the calculated stacking fault energy and the plasticity mechanisms in Fe-Mn-C alloys. *Mater. Sci. Eng. A* **2004**, *387–389*, 158–162. [CrossRef]
7. Linderov, M.; Segel, C.; Weidner, A.; Biermann, H.; Vinogradov, A. Deformation mechanisms in austenitic TRIP/TWIP steels at room and elevated temperature investigated by acoustic emission and scanning electron microscopy. *Mater. Sci. Eng. A* **2014**, *597*, 183–193. [CrossRef]
8. Pierce, D.T.; Jiménez, J.A.; Bentley, J.; Raabe, D.; Witting, J.E. The influence of stacking fault energy on the microstructural and strain-hardening evolution of Fe-Mn-Al-Si steels during tensile deformation. *Acta Mater.* **2015**, *100*, 178–190. [CrossRef]
9. Kim, J.K.; Chen, L.; Kim, H.S.; Kim, S.K.; Estrin, Y.; De Cooman, B.C. On the tensile behavior of high-manganese twinning-induced plasticity steel. *Metall. Mater. Trans. A* **2009**, *40*, 3147–3158. [CrossRef]
10. Chen, L.; Kim, H.S.; Kim, S.K.; De Cooman, B.C. Localized deformation due to Portevin-LeChatelier effect in 18Mn-0.6C TWIP austenitic steel. *ISIJ Int.* **2007**, *47*, 1804–1812. [CrossRef]
11. Lee, S.J.; Kim, J.Y.; Kane, S.N.; De Cooman, B.C. On the origin of dynamic strain aging in twinning-induced plasticity steels. *Acta Mater.* **2011**, *59*, 6809–6819. [CrossRef]
12. Kim, J.G.; Hong, S.; Anjabin, N.; Park, B.H.; Kim, S.K.; Chin, K.G.; Lee, S.; Kim, H.S. Dynamic strain aging of twinning-induced plasticity (TWIP) steel in tensile testing and deep drawing. *Mater. Sci. Eng. A* **2015**, *633*, 136–143. [CrossRef]
13. Taheri, A.K.; Maccagno, T.M.; Jonas, J.J. Dynamic strain aging and the wire drawing of low carbon steel rods. *ISIJ Int.* **1995**, *35*, 1532–1540. [CrossRef]
14. Li, C.C.; Leslie, W.C. Effects of dynamic strain aging on the subsequent mechanical properties of carbon steels. *Metall. Trans. A* **1978**, *9*, 1765–1775. [CrossRef]
15. Shahriary, M.S.; Koohbor, B.; Ahadi, K.; Ekrami, A.; Khakian-Qomi, M.; Izadyar, T. The effect of dynamic strain aging on room temperature mechanical properties of high martensite dual phase (HMDP) steel. *Mater. Sci. Eng. A* **2012**, *550*, 325–332. [CrossRef]
16. Bayramin, B.; Şimşir, C.; Efe, M. Dynamic strain aging in DP steels at forming relevant strain rates and temperatures. *Mater. Sci. Eng. A* **2017**, *704*, 164–172. [CrossRef]
17. Jung, I.C.; De Cooman, B.C. Temperature dependence of the flow stress of Fe-18Mn-0.6C-xAl twinning-induced plasticity steel. *Acta Mater.* **2013**, *61*, 6724–6735. [CrossRef]
18. Benzing, J.T.; Poling, W.A.; Pierce, D.T.; Witting, J.E. Effects of strain rate on mechanical properties and deformation behavior of an austenitic Fe-25Mn-3Al-3Si TWIP-TRIP steel. *Mater. Sci. Eng. A* **2018**, *711*, 78–92. [CrossRef]
19. Asghari, A.; Zarei-Hanzaki, A.; Eskandari, M. Temperature dependence of plastic deformation mechanisms in a modified transformation-twinning induced plasticity steel. *Mater. Sci. Eng. A* **2013**, *579*, 150–156. [CrossRef]
20. Saeed-Akbari, A.; Imlau, J.; Prahl, U.; Bleck, W. Derivation and variation in composition-dependent stacking fault energy maps based on subregular solution model in high-manganese steels. *Metall. Mater. Trans. A* **2009**, *40*, 3076–3090. [CrossRef]

21. Curtze, S.; Kuokkala, V.T. Dependence of tensile deformation behavior of TWIP steels on stacking fault energy, temperature and strain rate. *Acta Mater.* **2015**, *58*, 5129–5141. [CrossRef]

22. Allain, S.; Chateau, J.P.; Bouaziz, O. A physical model of the twinning-induced plasticity effect in a high manganese austenitic steel. *Mater. Sci. Eng. A* **2004**, *387–389*, 143–147. [CrossRef]

23. Renard, K.; Ryelandt, S.; Jacques, P.J. Characterisation of the Portevin-Le Châtelier effect affecting an austenitic TWIP steel based on digital image correlation. *Mater. Sci. Eng. A* **2010**, *527*, 2969–2977. [CrossRef]

24. Martin, S.; Wolf, S.; Martin, U.; Krüger, L.; Rafaja, D. Deformation mechanisms in austenitic TRIP/TWIP steel as a function of temperature. *Metall. Mater. Trans. A* **2016**, *47*, 49–58. [CrossRef]

25. Eskandari, M.; Zarei-Hanzaki, A.; Marandi, A. An investigation into the mechanical behavior of a new transformation-twinning induced plasticity steel. *Mater. Des.* **2012**, *39*, 279–284. [CrossRef]

26. Lan, P.; Zhang, J.Q. Twinning and dynamic strain aging behavior during tensile deformation of Fe-Mn-C TWIP steel. *Mater. Sci. Eng. A* **2017**, *700*, 250–258. [CrossRef]

27. Jacques, P.; Ladriere, J.; Delannay, F. On the influence of interactions between phases on the mechanical stability of retained austenite in transformation-induced plasticity multiphase steels. *Metall. Mater. Trans. A* **2001**, *32*, 2759–2768. [CrossRef]

28. Egea, A.J.S.; Rojas, H.A.G.; Celentano, D.J.; Peiró, J.J. Mechanical and metallurgical changes on 308L wires drawn by electropulses. *Mater. Des.* **2016**, *90*, 1159–1169. [CrossRef]

29. Lee, S.; Estrin, Y.; De Cooman, B.C. Effect of the strain rate on the TRIP-TWIP transition in austenitic Fe-12 pct Mn-0.6 pct C TWIP steel. *Metall. Mater. Trans. A* **2014**, *45*, 717–730. [CrossRef]

30. Van der Wegen, G.J.L.; Bronsveld, P.M.; Hosson, J.T.M. A comparison between different theories predicting the stacking fault energy from extended nodes. *Scr. Metall.* **1980**, *4*, 285–288. [CrossRef]

31. Ferreira, P.J.; Müllner, P. A thermodynamic model for the stacking-fault energy. *Acta Mater.* **1998**, *46*, 4479–4484. [CrossRef]

32. Peng, K.P.; Qian, K.W.; Chen, W.Z. Effect of dynamic strain aging on high temperature properties of austenitic stainless steel. *Mater. Sci. Eng. A* **2004**, *379*, 372–377. [CrossRef]

33. Molaei, M.J.; Ekrami, A. The effect of dynamic strain aging on fatigue properties of dual phase steels with different martensite morphology. *Mater. Sci. Eng. A* **2009**, *527*, 235–238. [CrossRef]

34. Bouaziz, O.; Allain, S.; Scott, C. Effect of grain and twin boundaries on the hardening mechanisms of twinning-induced plasticity steels. *Scr. Mater.* **2008**, *58*, 484–487. [CrossRef]

Article

Effect of Deformation Temperature on the Portevin-Le Chatelier Effect in Medium-Mn Steel

Barbara Grzegorczyk [1], Aleksandra Kozłowska [1], Mateusz Morawiec [1], Rafał Muszyński [2] and Adam Grajcar [1,*]

[1] Silesian University of Technology, Faculty of Mechanical Engineering, 18a Konarskiego Street, 44-100 Gliwice, Poland; barbara.grzegorczyk@polsl.pl (B.G.); aleksandra.kozlowska@polsl.pl (A.K.); mateusz.morawiec@polsl.pl (M.M.)

[2] Premco Logistics, 15/I/6 3 maja street, 41-200 Sosnowiec, Poland; rafal.muszynski@premco.eu

* Correspondence: adam.grajcar@polsl.pl; Tel.: +48-32-237-2933

Received: 24 November 2018; Accepted: 18 December 2018; Published: 20 December 2018

Abstract: Experimental investigations of the plastic instability phenomenon in a hot-rolled medium manganese steel were performed. The effects of tensile deformation in a temperature range of 20–140 °C on the microstructure, mechanical properties, and flow stress serrations were analyzed. The Portevin–Le Chatelier (PLC) phenomenon was observed for the specimens deformed at 60 °C, 100 °C, and 140 °C. It was found that the deformation temperature substantially affects the type and intensity of serrations. The type of serration was changed at different deformation temperatures. The phenomenon was not observed at room temperature. The plastic instability occurring for the steel deformed at 60 °C was detected for lower strain levels than for the specimens deformed at 100 °C and 140 °C. The increase of the deformation temperature to 100 °C and 140 °C results in shifting the PLC effect to a post uniform deformation range. The complex issues related to the interaction of work hardening, the transformation induced plasticity (TRIP) effect, and the PLC effect stimulated by the deformation temperature were addressed.

Keywords: medium-Mn steel; retained austenite; Portevin–Le Chatelier phenomenon; strain-induced martensitic transformation; ultra-high strength steel

---

1. Introduction

A growing interest in medium manganese steels related to their advantageous strength-ductility balance has prompted a better understanding of their behavior during plastic deformation [1–4]. Medium manganese steels contain 3–12% Mn and other alloying additions, such as Al and Si. In the present work, Mo addition is also added due to its strong solid solution strengthening potential [3]. Low-C (i.e., ~0.1%) steels containing ~3% Mn show ferritic-martensitic microstructures in the initial state [1]. When the manganese content rises, the initial microstructure (after cold rolling) changes to low-C martensite [2]. Medium-Mn steels can be obtained as cold-rolled [5] or hot-rolled [6] sheets. For lightweight applications in the automotive industry, cold-rolled coated sheet products are of particular interest. In order to obtain the multiphase microstructure consisting of ferrite and austenite or bainite and austenite, intercritical annealing is required [5,7]. These steels can be also produced as hot-rolled sheets for some underbody/suspension applications. After the hot-working and cooling to room temperature (a quenching step), the martensitic or ferritic-martensitic microstructure is subjected to intercritical annealing. The fine-grained lath-type ferrite-austenite mixture is formed under such annealing conditions [6]. For such small grain sizes of the austenite and its Mn contents higher than 7%, the stable austenite is retained at room temperature. For smaller Mn contents (i.e., for higher martensite start temperatures), an isothermal holding step at a bainitic range is required to be able to stabilize the retained austenite through its carbon enrichment [8]. An alternative approach is to apply thermomechanical processing (TMP) as direct one-step cooling following the hot rolling. The TMP has the potential for energy savings and high productivity due to the elimination of the need for successive heating [9–11]. Chemical composition and processing parameters are designed to obtain the optimal TRIP (Transformation Induced Plasticity) effect, which ensures superior mechanical properties [12,13].

Besides the many advantages of this group of steels, some problems during their processing can occur. Medium manganese steels [14–17] and high manganese steels [18–20] are prone to plastic instabilities phenomena associated with a serrated flow behavior–Portevin-Le Chatelier (PLC) effect and the appearance of Lüders bands. This is especially the case for cold-rolled steel sheets [21]. Moreover, it critically depends on the carbon level. With a higher C content, a pronounced effect is more visible. The heterogeneous deformation related to the increase in flow stress can lead to numerous cracks during the forming of sheets. Moreover, delayed cracking after deep drawing can appear. Delayed cracking is mostly observed in high-Mn content steels of the high austenite volume fraction. With the increased volume fraction of (retained) austenite, hydrogen embrittlement may be a more important concern. TWIP steels in the last decades have displayed important delayed cracking, which was industrially solved by decreasing the C content, Al alloying, and/or metallurgical processing [22,23]. Medium Mn steels have a less pronounced problem of delayed cracking compared to high-Mn steels [24,25].

The PLC effect is commonly known as characteristic serrations which can be observed on a strain-stress curve during a tensile test. The PLC effect has not been fully characterized yet. However, some correlations between the TRIP and PLC effects were recently documented in the literature. Gibbs et al. [26] proposed the theory that the characteristic serration is related to the various martensitic transformation rates during plastic deformation. Callahan et al. [14] observed that the TRIP effect and the passage of both Lüders and PLC bands occur simultaneously in 0.2C-5Mn-2.5Al medium-Mn steel. Sun et al. [15] reported that plastic instability in 0.26C-11.6Mn-2.7Al steel was related to discontinuous strain-induced martensite transformation. They observed localized martensite formation in the PLC bands, which propagated during tensile straining. It seems that twinning should also have some impact, taking into account the stacking fault energy of the alloy and its resulting increased austenite stability. Experimental results obtained by other authors [20,27,28] confirmed that the chemical composition of steel (especially C and Mn contents), deformation temperature, and strain rate had an influence on the occurrence of the PLC effect and the mechanical stability of the retained austenite.

So far, the PLC effect has been intensively studied in aluminium- and copper-based alloys [29–31]. Recently, this phenomenon was also largely studied in TWIP steels [18–20,22,23]. There are numerous theories explaining this effect. The most important factor is the interaction between dislocations and segregating C atoms. The occurrence of the PLC effect results in the appearance of technological problems during plastic forming. The characterization of factors affecting the PLC effect in medium-Mn steels is important both from a scientific point of view and their industrial application. Therefore, the goal of the current study is to identify the work hardening behavior of thermomechanically processed medium-Mn steel in terms of the PLC effect stimulated by a deformation temperature.

2. Material and Methods

2.1. Material

The chemical composition of the investigated steel is shown in Table 1. It is well-known [32,33] that an increased carbon content deteriorates the steel weldability. Moreover, the probability of the occurrence of the PLC effect becomes higher [19]. Hence, a relatively low carbon content of the investigated steel (0.16 wt.%) was designed. Manganese content also affects the PLC effect [34]. The tested steel contains ~5 wt.% Mn due to its application for hot-rolled products. The higher Mn contents are usually used for cold-rolled products [1,35]. Aluminium was chosen to prevent carbide precipitation.

The investigated steel was prepared by vacuum induction melting. Then, the laboratory ingots were hot-forged to obtain flat samples with a thickness of 22 mm. The next steps included roughing rolling and the thermomechanical rolling process.

Table 1. Chemical composition of the investigated medium-Mn steel in wt.%.

C %	Mn %	Al %	Si %	Mo %	P %	S %
0.16	4.7	1.6	0.22	0.20	0.008	0.004

2.2. Thermomechanical Rolling

The thermomechanical laboratory rolling included three passes. The final sheet thickness was about 4.5 mm. Figure 1 shows the applied hot-rolling conditions. The austenitization was carried out at 1200 °C. The temperature of the final pass was about 850 °C. Then, the flat steel samples were cooled to 400 °C and held at this temperature for 300 s. Finally, air cooling to room temperature was applied.

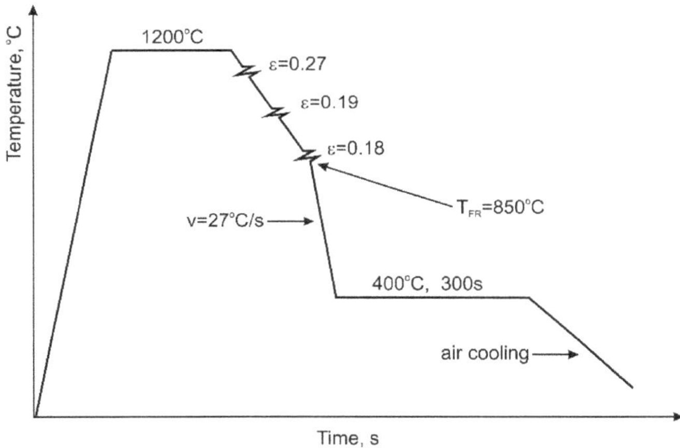

Figure 1. Parameters of thermomechanical processing of investigated steel.

2.3. Experimental Procedures

The tensile tests were performed at 20 °C and at elevated temperatures: 60 °C, 100 °C, and 140 °C, to determine the effect of deformation temperature on the PLC effect and the related microstructure and mechanical properties. Tensile specimens were machined from the hot-rolled sheet along the rolling direction. Three uniaxial tensile tests were carried out for each temperature using an INSTRON 1195 universal testing machine (INSTRON, Norwood, MA, USA)at a strain rate of 10^{-3} s^{-1}. The results of tensile tests allowed us to determine the effect of deformation temperature on the mechanical properties. A determination of the work hardening rate allowed us to establish the influence of the TRIP effect on the strengthening behavior. Based on σ—ε curves, classification of the strengthening stages and types of oscillations was conducted. The value of the critical strain, required for the initiation of the PLC effect, was also determined.

The microstructural analysis was performed for the steel samples tensile tested at different temperatures. The observations were carried out using the Zeiss Axio Observer Z1m optical microscope (Carl Zeiss AG, Jena, Germany). The microstructural details were revealed with a scanning electron microscope Zeiss SUPRA 25 (Carl Zeiss AG, Jena, Germany) operating at 20 kV, working in SE mode. Specimens for microstructural observations were prepared using standard metallographic procedures. They were mechanically ground with SiC paper up to 2000 grid, then polished with a diamond paste and finally etched in 5% nital.

3. Results

3.1. Mechanical Behavior

Results of static tensile tests (average values of three samples) performed at various temperatures: 20 °C, 60 °C, 100 °C, and 140 °C, are listed in Table 2 and presented in Figure 2.

Table 2. Mechanical properties of the investigated steel.

Deformation Temperature [°C]	UTS [MPa]	YS [MPa]	TE [%]	YS/UTS [MPa]
20	1215	882	11.5	0.73
60	1200	856	10.3	0.71
100	1180	873	9.8	0.74
140	1223	841	11.4	0.69

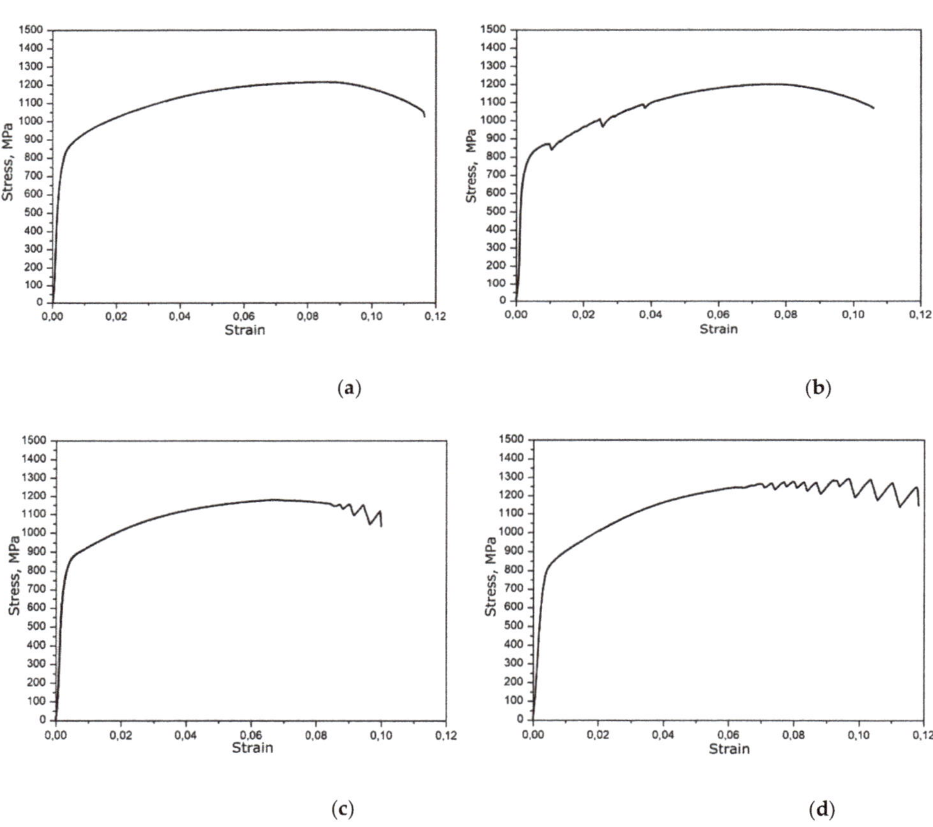

Figure 2. Engineering stress-strain curves of investigated steel registered at: (**a**) 20 °C, (**b**) 60 °C, (**c**) 100 °C, and (**d**) 140 °C.

Based on the obtained data, a pronounced relationship between the deformation temperature and mechanical properties was observed. The sample deformed at 20 °C shows relatively high mechanical properties. The tensile strength (UTS) and yield stress (YS) values are 1215 MPa and 882 MPa, respectively (Table 2). The highest total elongation (TE) value was obtained at this deformation temperature −11.5%. The YS/UTS ratio is 0.73, which indicates the relatively high hardening potential. Increasing the deformation temperature to 60 °C results in slight lowering of both UTS and YS values

(UTS = 1200 MPa; YS = 856 MPa). The TE value is also a little bit lower (10.3%). The YS/UTS ratio is quite similar to that obtained at the lower deformation temperature. At 100 °C, a similar yield stress is obtained. The tensile strength (1180 MPa) is slightly lower as a result of smaller work hardening. A further deterioration of TE was detected (9.8%). Surprisingly, the highest UTS value (1223 MPa) was obtained for the specimen deformed at 140 °C. The YS/UTS ratio in this case is 0.69, which indicates the relatively high strengthening potential. As a result, the TE value was about 11.4%, which is similar to the TE value obtained for the sample deformed at 20 °C (Table 2).

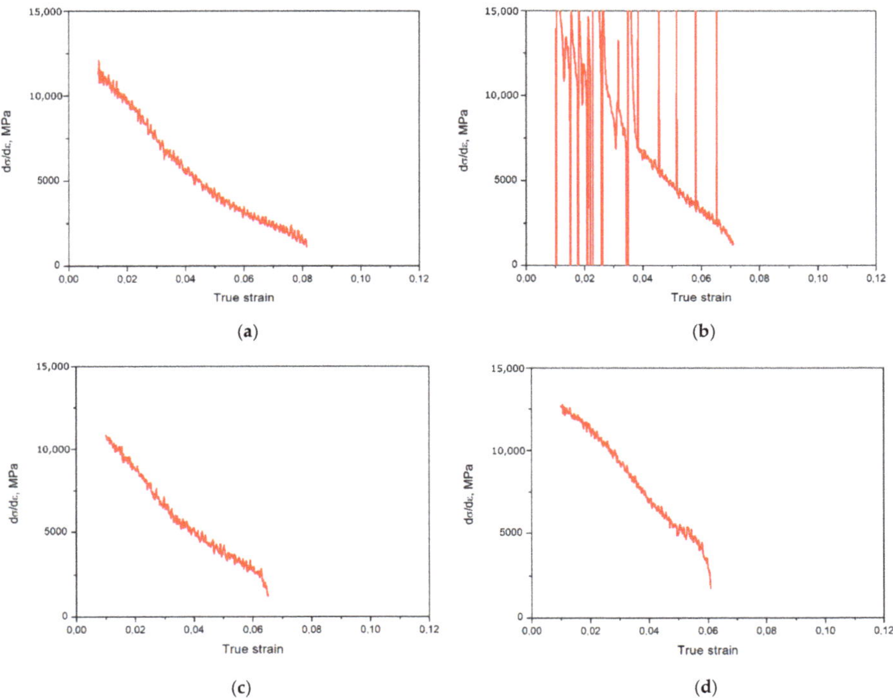

Figure 3. Work hardening rate as a function of true strain (uniform deformation range) at: (**a**) 20 °C, (**b**) 60 °C, (**c**) 100 °C, and (**d**) 140 °C.

Some typical features for the PLC effect [14–20,28] were observed for the investigated steel deformed at elevated temperatures: 60–140 °C. The characteristic serrations were observed on the stress-strain curves (Figure 2). The specimen deformed at 20 °C showed no evidence of the PLC effect (Figure 2a). Increasing the deformation temperature to 60 °C resulted in the appearance of serrations in the initial stage of plastic deformation (Figure 2b). Observed oscillations were classified as A type according to the nomenclature given in previous works [36,37]. The stress oscillation occurs periodically on the graph line after the yield point. The range of deformation when serration occurred is variable. In contrast, for the specimens deformed at 100 °C and 140 °C, the oscillations were observed in the final stage of deformation (Figure 2c,d). The serrations were observed in the range of post-uniform deformation. The shape of the serrations at 100 °C corresponds to the type B [36]. Increasing the deformation temperature to 140 °C results in intensifying the serration amplitude and plastic deformation range. This occurred in the whole range of post-uniform elongation (Figure 2d). After the strain of ca. 0.08, the B+C type oscillations [37] can be observed. The corresponding increase of TE to 11.4% should be related to the enhanced occurrence of oscillations due to the PLC effect. Generally, the important influence of the deformation temperature on the oscillation type and its intensity was observed. In the literature [14,15,17], the mechanism of plastic instability phenomenon associated with the Portevin-Le Chatelier effect in Fe-Mn alloys at an elevated temperature is generally attributed to a dynamic strain aging (DSA) effect.

The deformation temperature strongly affects the work hardening rate of the investigated steel (Figure 3). The $d\sigma/d\varepsilon$ values decrease gradually as the deformation degree increases. This behavior is typical. In steels with a high volume fraction of metastable retained austenite, a local increase in the $d\sigma/d\varepsilon$ value can be observed [2,13] due to the intense TRIP effect. This is not the case for the present steels. The smallest slope of the curves was observed for the steel deformed at 20 °C (Figure 3a). When increasing the deformation temperature from 20 °C to 140 °C, a general slope of the $d\sigma/d\varepsilon$-ε curves rises. It is consistent with the gradual decrease of true strain in a range of uniform deformation: from 0.08 at 20 °C to 0.06 at 140 °C. The different course of the curve in Figure 3b is related to the appearance of the PLC effect in the early stages of uniform strain. In this case, the work hardening rate is strongly irregular due to the serrations occurring in the tensile curve. The lowest work hardening rate was detected for the steel deformed at 100 °C (11000 MPa). The work hardening rate was similar for the specimens deformed at 20 °C and 140 °C. This means that some thermally activated ageing processes are enhanced after the rise in temperature to 140 °C.

Mechanical properties of the steels were also assessed by hardness tests using the Vickers method. The obtained results are presented in Figure 4. It is clear that hardness values are affected by the deformation temperature. They are at a relatively high level taking into account the carbon content in the steel (490–530 HV). The lowest hardness was detected for the specimen in a hot-rolled state. All the hardness results were higher for plastically deformed samples. It was found that the highest hardness was obtained for the sample tensile tested at room temperature. Then, the hardness decreases when increasing the deformation temperature. The presented results are in reasonable agreement with the mechanical properties obtained from the static tensile tests (Table 2). Some variations could be due to the different intensity of the TRIP effect and PLC effect at different temperatures.

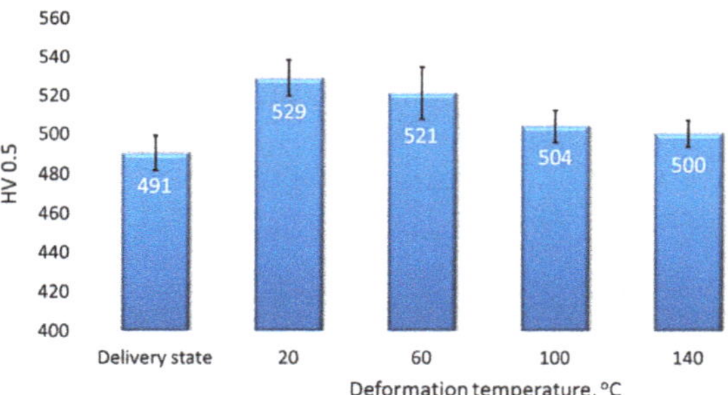

Figure 4. Effect of deformation temperature on the hardness.

3.2. Microstructure

Results of metallographic observations allowed us to determine the microstructure of the investigated steel in the delivery state (after hot rolling) and after the tensile test. Optical micrographs for selected deformation temperatures are shown in Figure 5. The hot-rolled 5Mn steel (in the delivery state) is characterized by a mixture of bainite and martensite (Figure 5a). At this magnification, it is hard to assess the presence of retained austenite. However, the amount of this phase determined in our previous works [8,38] by XRD methods is equal to ~10%. The microstructures registered after the plastic deformation in a temperature range of 20–140 °C are very similar. Fine laths of bainitic-martensitic products are visible (Figure 5b–d).

The identification of retained austenite in the microstructure of 5Mn steel is difficult due to its relatively small amount. It is related to the manganese content (5 wt.%), which affects the carbon content level in the retained austenite. As the manganese content increases, the carbon content in the γ phase decreases [39–41]. This was recently confirmed by Sugimoto et. al [1] in medium-Mn steels containing 1.5 to 5% Mn.

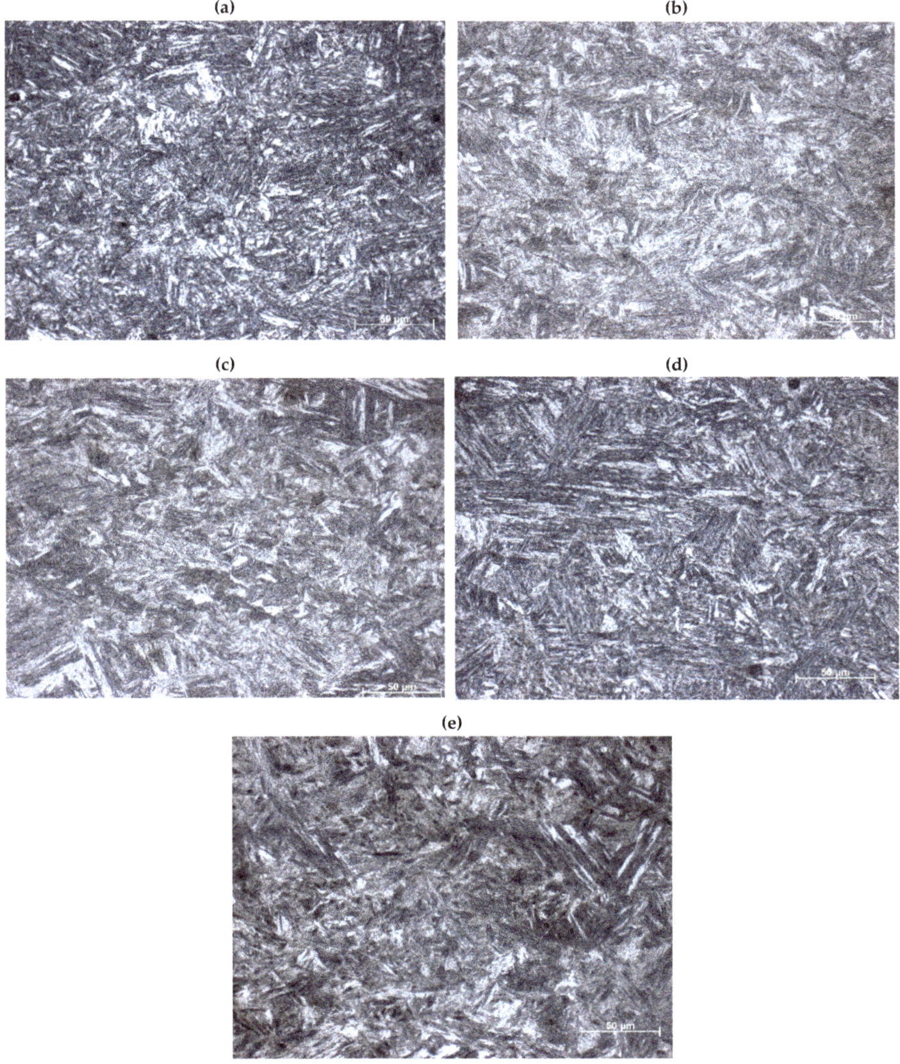

Figure 5. Optical micrographs of the investigated steel: (**a**) delivery state; deformed at: (**b**) 20 °C, (**c**) 60 °C, (**d**) 100 °C, and (**e**) 140 °C.

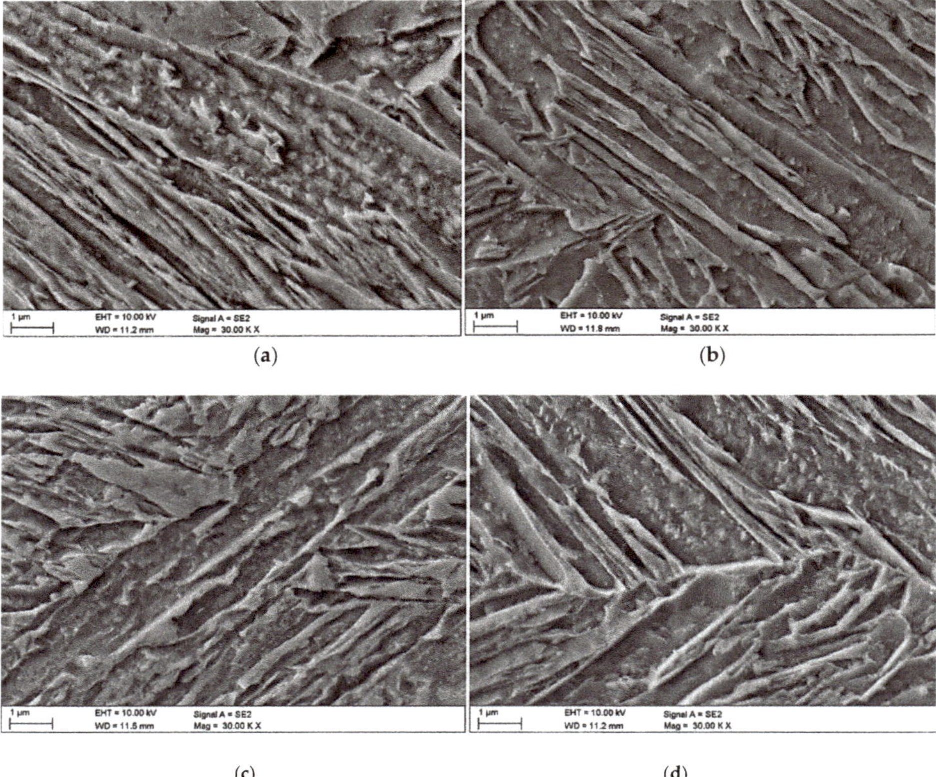

Figure 6. SEM micrographs of the 5Mn steel deformed at: (**a**) 20 °C, (**b**) 60 °C, (**c**) 100 °C, and (**d**) 140 °C.

SEM observations confirmed the presence of the lath-type martensitic-bainitic microstructure (Figure 6). Bright, thin layers observed in the microstructure are probably the remains of retained austenite [8,38]. This metastable phase is partially transformed into strain-induced martensite. However, this requires more detailed microscopic investigations. The laths have various thicknesses. Inside the bainitic areas, small precipitates characterized by a different shape can also be identified. This indicates the initiation of carbide precipitation during the isothermal holding step following the hot rolling.

4. Discussion

Plastic deformation of medium manganese steels with a multiphase microstructure is very complex. The deformation of these steels is related to the specific microstructural mechanisms, like transformation induced plasticity (TRIP) and Portevin-Le Chatelier (PLC) effects. The occurrence of the PLC effect was previously typically observed in Al and Cu alloys at room or elevated temperatures [29–31]. Concerning steels, the serrated behavior in flow stress was observed in dual-phase steels [42], quenching and partitioning steels [28], and recently in high manganese twinning induced plasticity (TWIP) steels containing 17-25 % Mn [18–20,23,27,43]. Increasing the Mn content reduces the diffusion rate of carbon [44]. That is why the PLC effect is so pronounced in TWIP steels. However, recent studies proved that it also takes place in medium-Mn steels [15,24,25]. This process usually takes place in a range of specific temperatures and strain rates. An explanation of the PLC phenomenon in medium manganese steels seems to be a difficult issue due to the multiphase microstructure and the corresponding occurrence of the TRIP effect during plastic deformation. The temperature effect makes the analysis more advanced.

So far, in the literature, there are only a few reports available concerning the PLC effect in medium manganese steels [14–17,21,24]. Most of them concern cold rolled-steels. The PLC effect in hot-rolled medium Mn steels has rarely been analyzed [25]. Based on the data available in the literature and results obtained in the present study, one can conclude that the first difference exists in the yield point. In contrast to cold-rolled and intercritically annealed steels, the thermomechanically processed steels did not show evidence of high and low yield stress and subsequent Lüders elongation. This is the case for intercritically annealed steels [14,16] and can be a real industrial problem during stamping, etc. The present steels show a continuous yielding behavior as a result of the mobile dislocations generated during previous thermomechanical processing. This was shown in our previous works [8,38].

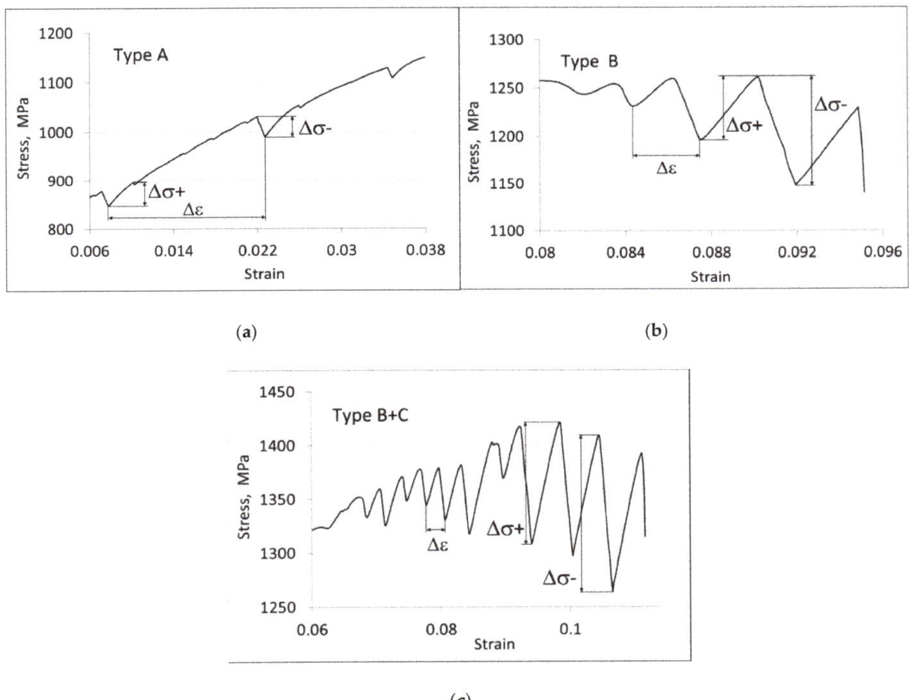

Figure 7. Details of serrations at the flow curves for the steel deformed at: 60 °C (**a**), 100 °C (**b**), and 140 °C (**c**).

Results obtained in the present study showed that the temperature factor has a significant influence on the PLC occurrence. The steel deformed at 20 °C showed no evidence of plastic instability (Figure 2a). Characteristic serrations were detected at elevated deformation temperatures: 60 °C, 100 °C, and 140 °C (Figure 2b–d). It is interesting that the type of oscillations and their amplitude substantially depend on the deformation temperature. The specimen deformed at 60 °C showed oscillation classified as an A type. The serration occurred in a uniform elongation range, i.e., earlier than in steels deformed at higher temperatures—already at a strain level of 0.006. Moreover, this appeared regularly and its amplitude was quite low (Figure 7a). A similar type of serration was observed in medium-Mn steels by other authors [14,15]. However, the type of serration and plastic instability deformation range shifted to a higher deformation level when increasing the deformation temperature. The instable flow behavior was initiated in a post-uniform elongation range. Moreover, the amplitude of serration rose as the deformation temperature increased (Figure 7b,c). Sun et al. [15] reported that this can be due to the higher fraction of untransformed austenite at higher deformation temperatures. This phase is especially sensitive to the plastic instability phenomenon, like in pure high-Mn steels [18–20]. The type of serration observed at 100 °C was classified as a type B. In the case of 140 °C, the B and C types occurred with the highest amplitude and strain extent (Table 3).

Table 3. Detailed characteristics of serration flow at various deformation temperatures.

Deformation Temperature [°C]	Serration Type	ε_{cr}	$\Delta\sigma+_{max}$ [MPa]	$\Delta\sigma-_{max}$ [MPa]	$\overline{\Delta\sigma+}$ [MPa]	$\overline{\Delta\sigma-}$ [MPa]	$\overline{\Delta\varepsilon}$
60	A	0.007	79	41	45	14	0.005
100	B	0.080	82	114	46	67	0.003
140	B+C	0.060	129	126	68	64	0.004

ε_{cr}—critical strain for initiation of the PLC effect
$\Delta\sigma+_{max}$—maximum increase of the oscillation stress
$\Delta\sigma-_{max}$—maximum decrease of the oscillation stress
$\overline{\Delta\sigma+}$—mean value of the stress increase in a range of oscillations
$\overline{\Delta\sigma-}$—mean value of the stress decrease in a range of oscillations
$\overline{\Delta\varepsilon}$—mean value of the strain period between successive oscillations

Based on the data obtained by other authors [28,42], it seems that the deformation temperatures applied in the present study (60–140 °C) are sufficient to induce diffusion processes of solutes. As a result, it leads to the appearance of dynamic strain ageing (DSA). In general, the DSA arises from the dynamic interactions between mobile dislocations and solute atoms [15]. Jung and De Cooman [45] reported that in medium-Mn steels, this effect results from an interaction between dislocations and Mn-C clusters. They suggested that the movement of C atoms in Mn-C clusters between octahedral and tetrahedral interstices could lock partial dislocations in stacking faults. Hickel et al. [46] also confirmed a strong effect of Mn-C clusters on the SFE and showed that the active deformation mechanism can locally redistribute solutes and thus significantly change the local SFE. These findings concern high-Mn steels and typical medium-Mn steels covering an Mn level from 7 to 12%. Moreover, the serration behavior is these steels already takes place at room temperature. In the present steel, the Mn content is limited to 5%. This is presumably why the PLC effect is shifted to a post-uniform strain range and this requires at least 100 °C. The serration is intensified at the higher temperature (140 °C) with the faster carbon diffusion.

Generally, when increasing the deformation temperature, the mechanical properties of medium manganese decrease. This is related to the lower contribution of the TRIP effect, i.e., the mechanical stability of retained austenite increases [47,48]. Since the retained austenite amount in the present steel is relatively low (~10%), the contribution of the TRIP effect to the PLC effect cannot be significant. This is especially true at 140 °C with the lowest driving force for the strain-induced martensitic transformation [26]. This is because the stacking fault energy increases when the temperature rises [20]. Hence, the most intense PLC effect occurred at that temperature, at which the highest volume fraction of retained austenite remains mechanically stable. Finally, it resulted in a slight increase of ductility of the sample deformed at 140 °C (Table 2). Sachdev [42] reported that martensite tempering at this temperature range can also contribute to some improvement of the ductility in steels containing some martensite fraction. This phenomenon makes the work hardening and corresponding flow serration analyses more complex, and more detailed work is thus required.

5. Conclusions

The present study addressed the effect of deformation temperature on the work hardening behavior and corresponding PLC effect in thermomechanically processed medium-Mn steel with 1.6% Al content. The steel is characterized by a bainitic-martensitic microstructure with some fraction of retained austenite. Tensile curves show the continuous yielding behavior without the yield point phenomenon and Lüders elongation. Tensile strength reaches 1200 MPa and ductility is limited to 10%. The mechanical properties are slightly temperature-dependent, whereas the temperatures factor affects the flow behavior significantly. The flow stress serration is activated at 60 °C in an initial range of uniform deformation. The instability of plastic deformation is postponed to a post-uniform deformation range in a temperature range of 100-140 °C. The highest amplitude and range of serrations occurred at 140 °C due to the highest mechanically stable fraction of retained austenite. The delay of the serration to the post-uniform strain range is due to the Mn content being limited to 5% compared to typical medium-Mn steels.

Author Contributions: Conceptualization, A.G.; Data curation, B.G., M.M. and R.M.; Formal analysis, A.G.; Investigation, B.G., A.K., M.M., R.M. and A.G.; Methodology, B.G., A.K. and R.M.; Project administration, A.G.; Visualization, M.M.; Writing—original draft, A.K.; Writing—review & editing, A.G.

Funding: The financial support of the National Science Center, Poland, is gratefully acknowledged, grant no. 2017/27/B/ST8/02864.

Acknowledgments: The work was supported by statutory funds from the Faculty of Mechanical Engineering of Silesian University of Technology in 2018.

Conflicts of Interest: The authors declare no conflict of interest.

References

1. Sugimoto, K.; Tanino, H.; Kobayashi, J. Impact toughness of medium-Mn transformation-induced plasticity-aided steels. *Steel Res. Int.* **2015**, *86*, 1151–1160. [CrossRef]
2. Steineder, K.; Krizan, D.; Schneider, R.; Beal, C.; Sommitsch, C. On the microstructural characteristics influencing the yielding behavior of ultra-fine grained medium-Mn steels. *Acta Mater.* **2017**, *139*, 39–50. [CrossRef]
3. Marcisz, J.; Stępień, J. Short-time ageing of MS350 maraging steel with and without plastic deformation. *Arch. Metall. Mater.* **2014**, *59*, 513–520. [CrossRef]
4. Grajcar, A.; Radwanski, K. Microstructural comparison of the thermomechanically treated and cold deformed Nb-microalloyed TRIP steel. *Mater. Tehnol.* **2014**, *48*, 679–683.
5. Li, X.; Song, R.; Zhou, N.; Li, J. An ultrahigh strength and enhanced ductility cold-rolled medium-Mn steel treated by intercritical annealing. *Scr. Mater.* **2018**, *154*, 30–33. [CrossRef]
6. Li, Z.C.; Ding, H.; Cai, Z.H. Mechanical properties and austenite stability in hot-rolled 0.2C–1.6/3.2Al–6Mn–Fe TRIP steel. *Mater. Sci. Eng. A* **2015**, *639*, 559–566. [CrossRef]
7. Mishra, G.; Chandan, A.K.; Kundu, S. Hot rolled and cold rolled medium manganese steel: Mechanical properties and microstructure. *Mater. Sci. Eng. A* **2017**, *701*, 319–327. [CrossRef]
8. Grajcar, A.; Kilarski, A.; Kozłowska, A. Microstructure-property relationships in thermomechanically processed medium-Mn steels with increased Al content. *Metals* **2018**, *8*. [CrossRef]
9. Opiela, M. Thermomechanical treatment of Ti-Nb-V-B micro-alloyed steel forgings. *Mater. Tehnol.* **2014**, *48*, 587–591.
10. Steineder, K.; Beal, C.; Krizan, D.; Dikovits, M.; Sommitsch, C.; Schneider, R. Hot deformation behavior of a 3rd generation advanced high strength steel with a medium-Mn content. *Key Eng. Mater.* **2015**, *651–653*, 120–125. [CrossRef]
11. Opiela, M.; Grajcar, A. Hot deformation behavior and softening kinetics of Ti-V-B microalloyed steels. *Arch. Civil Mechan. Eng.* **2012**, *12*, 327–333. [CrossRef]
12. Sugimoto, K.; Hidaka, S.; Tanino, H.; Kobayashi, J. Warm Formability of 0.2 Pct C-1.5 Pct Si-5 Pct Mn Transformation-Induced Plasticity-aided steel. *Metall. Mater. Trans. A Phys. Metall. Mater. Sci.* **2017**, *48*, 2237–2246. [CrossRef]
13. Chiang, J.; Lawrence, B.; Boyd, J.D.; Pilkey, A.K. Effect of microstructure on retained austenite stability and work hardening of TRIP steels. *Mater. Sci. Eng. A* **2011**, *528*, 4516–4521. [CrossRef]
14. Callahan, M.; Hubert, O.; Hild, F.; Perlade, A.; Schmitt, J.H. Coincidence of strain-induced TRIP and propagative PLC bands in medium Mn steels. *Mater. Sci. Eng. A* **2017**, *703*, 391–400. [CrossRef]

15. Sun, B.; Vanderesse, N.; Fazeli, F.; Scott, C.; Chen, J.; Bocher, P.; Jahazi, M.; Yue, S. Discontinuous strain-induced martensite transformation related to the Portevin-Le Chatelier effect in a medium manganese steel. *Scr. Mater.* **2017**, *133*, 9–13. [CrossRef]

16. Yang, F.; Luo, H.; Pu, E.; Zhang, S.; Dong, H. On the characteristics of Portevin–Le Chatelier bands in cold-rolled 7Mn steel showing transformation-induced plasticity. *Int. J. Plast.* **2018**, *103*, 188–202. [CrossRef]

17. Wang, X.G.; Wang, L.; Huang, M.X. Kinematic and thermal characteristics of Luders and Portevin-Le Chatelier bands in a medium Mn transformation-induced plasticity steel. *Acta Mater.* **2017**, *124*, 17–29. [CrossRef]

18. Allain, S.; Cugy, P.; Scott, C.; Chateau, J.P.; Rusinek, A.; Deschamps, A. The influence of plastic instabilities on the mechanical properties of a high-manganese austenitic FeMnC steel. *Int. J. Mater. Res.* **2008**, *99*. [CrossRef]

19. Chen, L.; Kim, H.S.; Kim, S.K.; De Cooman, B.C. Localized deformation due to Portevin–LeChatelier effect in 18Mn–0.6C TWIP austenitic steel. *ISIJ Int.* **2007**, *47*, 1804–1812. [CrossRef]

20. Lee, S.Y.; Lee, S.I.; Hwang, B. Effect of strain rate on tensile and serration behaviors of an austenitic Fe-22Mn-0.7C twinning-induced plasticity steel. *Mater. Sci. Eng. A* **2018**, *711*, 22–28. [CrossRef]

21. Wang, H.; Zhang, Y.; Yuan, G.; Kang, J.; Wang, Y.; Misra, R.D.K.; Wang, G. Significance of cold rolling reduction on Lüders band formation and mechanical behavior in cold-rolled intercritically annealed medium-Mn steel. *Mater. Sci. Eng. A* **2018**, *737*, 176–181. [CrossRef]

22. Kim, Y.; Kang, N.; Park, Y.; Choi, I.; Kim, G.; Kim, S.; Cho, K. Effects of the strain induced martensite transformation on the delayed fracture for Al-added TWIP steel. *J. Korean Inst. Met. Mater.* **2008**, *46*, 780–787.

23. Kusakin, P.S.; Kaibyshev, R.O. High-Mn Twinning Induced Plasticity Steels: Microstructure and mechanical properties. *Rev. Adv. Mater. Sci.* **2016**, *44*, 326–360.

24. Field, D.M.; Van Aken, D.C. Dynamic strain ageing phenomena and tensile response of medium-Mn TRIP steel. *Metall. Mater. Trans. A* **2018**, *49A*, 1152–1166. [CrossRef]

25. Hu, B.; Luo, H. Microstructures and mechanical properties of 7Mn steel manufactured by different rolling processes. *Metals* **2017**, *7*. [CrossRef]

26. Gibbs, P.J.; De Moor, E.; Merwin, M.J.; Clausen, B.; Speer, J.G.; Matlock, D.K. Austenite stability effects on tensile behavior of manganese-enriched-austenite transformation-induced plasticity steel. *Metall. Mater. Trans. A* **2011**, *42*, 3691–3702. [CrossRef]

27. Lee, S.J.; Kim, J.; Kane, S.N.; De Cooman, B.C. On the origin of dynamic strain aging in twinning-induced plasticity steels. *Acta Mater.* **2011**, *59*, 6809–6819. [CrossRef]

28. Min, J.; Hector, L.G., Jr.; Zhang, L.; Sund, L.; Carsley, J.E.; Lin, J. Plastic instability at elevated temperatures in a TRIP-assisted steel. *Mater. Des.* **2016**, *95*, 370–386. [CrossRef]

29. Ozgowicz, W.; Grzegorczyk, B.; Pawełek, A.; Wajda, W.; Skuza, W.; Piątkowski, A.; Ranachowski, Z. An analysis of the Portevin-Le Chatelier effect and cracking of CuSn6P alloy at elevated temperature of deformation applying the acoustic emission method. *Eng. Fract. Mechan.* **2016**, *167*, 112–122. [CrossRef]

30. Meng, X.; Liu, B.; Luo, L.; Ding, Y.; Rao, X.X.; Hu, B.; Liu, Y.; Lu, J. The Portevin-Le Chatelier effect of gradient nanostructured 5182 aluminum alloy by surface mechanical attrition treatment. *J. Mater. Sci. Technol.* **2018**, *34*, 2307–2315. [CrossRef]

31. Ozgowicz, W.; Grzegorczyk, B.; Pawełek, A.; Piątkowski, A.; Ranachowski, Z. Influence of the strain rate on the PLC effect and acoustic emission in single crystals of the CuZn30 alloy compressed at an elevated temperature. *Mater. Tehnol.* **2015**, *49*, 197–202. [CrossRef]

32. Górka, J. Welding thermal cycle-triggered precipitation processes in steel S700MC subjected to the thermo-mechanical control processing. *Arch. Metall. Mater.* **2017**, *62*, 321–326. [CrossRef]

33. Kurc-Lisiecka, A.; Piwnik, J.; Lisiecki, A. Laser welding of new grade of advanced high strength steel strenx 1100 MC. *Arch. Metall. Mater.* **2017**, *62*, 1651–1657. [CrossRef]

34. Sun, B.; Fazeli, F.; Scott, C.; Guo, B.; Aranas, C., Jr.; Chu, X.; Jahazi, M.; Yue, S. Microstructural characteristics and tensile behavior of medium manganese steels with different manganese additions. *Mater. Sci. Eng. A* **2018**, *729*, 496–507. [CrossRef]

35. Kamoutsi, H.; Gioti, E.; Haidemenopoulos, G.N.; Cai, Z.; Ding, H. Kinetics of solute partitioning during intercritical annealing of a medium-Mn steel. *Metall. Mater. Trans. A* **2015**, *46*, 4841–4846. [CrossRef]

36. Cottrell, A.H. A note on the Portevin-le Chatelier effect. *Lond. Edinb. Dublin Philos. Mag. J. Sci.* **1953**, *44*, 829–832. [CrossRef]

37. Mc Cormick, P.G. Theory of flow localization due to dynamic strain ageing. *Acta Mater.* **1988**, *36*, 3061–3067. [CrossRef]
38. Grajcar, A.; Skrzypczyk, P.; Kozłowska, A. Effects of temperature and time of isothermal holding on retained austenite stability in medium-Mn steels. *Appl. Sci.* **2018**, *8*. [CrossRef]
39. Gomez, M.; Garcia, C.I.; De Ardo, A.J. The role of new ferrite on retained austenite stabilization in Al-TRIP steels. *ISIJ Int.* **2010**, *50*, 139–146. [CrossRef]
40. Jacques, P.J.; Girault, E.; Mertens, A.; Verlinden, B.; Van Humbeeck, J.; Delannay, F. The developments of cold-rolled TRIP-assisted multiphase steels. Al-alloyed TRIP-assisted multiphase steels. *ISIJ Int.* **2001**, *41*, 1068–1074. [CrossRef]
41. Soliman, M.; Palkowski, H. On factors affecting the phase transformation and mechanical properties of cold-rolled transformation-induced-plasticity-aided steel. *Metall. Mater. Trans. A* **2008**, *39*, 2513–2527. [CrossRef]
42. Sachdev, A.K. Dynamic strain aging of various steels. *Metall. Trans. A* **1982**, *13*, 1793–1797. [CrossRef]
43. Aydemir, B.; Zeytin, H.K.; Guven, G. Investigation of Portevin-Le Chatelier effect of hot-rolled Fe-13Mn-0.2C-1Al-1Si TWIP steel. *Mater. Tehnol.* **2016**, *50*, 511–516. [CrossRef]
44. Kral, L.; Million, B.; Cermak, J. Diffusion of carbon and manganese in Fe-Mn-C. *Defect Diffus. Forum* **2007**, *263*, 153–158. [CrossRef]
45. Jung, I.C.; De Cooman, B.C. Temperature dependence of the flow stress of Fe-18Mn-0.6C-xAl twinning-induced plasticity steel. *Acta Mater* **2013**, *61*, 6724–6735. [CrossRef]
46. Hickel, T.; Sandlöbes, S.; Marceau, R.K.W.; Dick, A.; Bleskov, I.; Neugebauer, J.; Raabe, D. Impact of nanodiffusion on the stacking fault energy in high-strength steels. *Acta Mater.* **2014**, *75*, 147–155. [CrossRef]
47. Kim, H.; Lee, J.; Barlat, F.; Kim, D.; Lee, M.G. Experiment and modeling to investigate the effect of stress state, strain and temperature on martensitic phase transformation in TRIP-assisted steel. *Acta Mater.* **2015**, *97*, 435–444. [CrossRef]
48. Zhang, M.; Li, L.; Ding, J.; Wu, Q.; Wang, Y.D.; Almer, J.; Guo, F.; Ren, Y. Temperature-dependent micromechanical behavior of medium-Mn transformation-induced-plasticity steel studied by in situ synchrotron X-ray diffraction. *Acta Mater.* **2017**, *141*, 294–303. [CrossRef]

Article

Effect of Manganese on the Structure-Properties Relationship of Cold Rolled AHSS Treated by a Quenching and Partitioning Process

Simone Kaar [1],*, Daniel Krizan [2], Reinhold Schneider [1], Coline Béal [3] and Christof Sommitsch [3]

[1] Research and Development, University of Applied Sciences Upper Austria, Wels 4600, Austria; reinhold.schneider@fh-wels.at

[2] Research and Development Department, Business Unit Coil, voestalpine Stahl GmbH, Linz 4020, Austria; daniel.krizan@voestalpine.com

[3] Institute of Materials Science, Joining and Forming, Graz University of Technology, Graz 8010, Austria; coline.beal@tugraz.at (C.B.); christof.sommitsch@tugraz.at (C.S.)

* Correspondence: simone.kaar@fh-wels.at; Tel.: +43-50304-15-6250

Received: 17 September 2019; Accepted: 16 October 2019; Published: 19 October 2019

Abstract: The present work focuses on the investigation of both microstructure and resulting mechanical properties of different lean medium Mn Quenching and Partitioning (Q&P) steels with 0.2 wt.% C, 1.5 wt.% Si, and 3–4 wt.% Mn. By means of dilatometry, a significant influence of the Mn-content on their transformation behavior was observed. Light optical and scanning electron microscopy (LOM, SEM) was used to characterize the microstructure consisting of tempered martensite (α''), retained austenite (RA), partially bainitic ferrite (α_B), and final martensite (α'_{final}) formed during final cooling to room temperature (RT). Using the saturation magnetization measurements (SMM), a beneficial impact of the increasing Mn-content on the volume fraction of RA could be found. This remarkably determined the mechanical properties of the investigated steels, since the larger amount of RA with its lower chemical stabilization against the strain-induced martensite transformation (SIMT) highly influenced their overall stress-strain behavior. With increasing Mn-content the ultimate tensile strength (UTS) rose without considerable deterioration in total elongation (TE), leading to an enhanced combination of strength and ductility with UTS × TE exceeding 22,500 MPa%. However, for the steel grades containing an elevated Mn-content, a narrower process window was observed due to the tendency to form α'_{final}.

Keywords: lean medium Mn Q&P steel; stress-strain behavior; mechanical properties; retained austenite stability

1. Introduction

Increasing requirements of the automotive industry related to lightweight construction and increased passenger safety drive the development of advanced high strength steels (AHSS) [1,2]. Currently, research focuses on the development of the third generation AHSS, including the concepts of medium Mn and Quenching and Partitioning (Q&P) steels. These steel grades offer a promising combination of strength and ductility achieved by a microstructure having a substantial amount of retained austenite (RA) which transforms into strain-induced martensite (α') due to the transformation induced plasticity (TRIP) effect [3–5].

Medium Mn steels with a typical chemical composition of 0.05–0.2 wt.% C and 3–10 wt.% Mn have a microstructure consisting of an ultrafine-grained ferritic (α) matrix and volume fractions of retained austenite (RA) up to 40 vol.%. Therefore, they are characterized by an excellent combination of strength and ductility, achieving ultimate tensile strengths (UTS) > 800 MPa combined with total elongations (TE) of up to 40% [6–8].

Q&P is being considered as a novel heat treatment to produce steels with a carbon-depleted α'' matrix that contain a considerable volume fraction of RA. The Q&P process was first proposed by Speer et al. [9], and consists of a two-step heat treatment. After heating in order to obtain a fully austenitic microstructure, the steel is initially quenched to a specific quenching temperature (T_Q) in the M_S-M_f temperature range, where austenite partially transforms into primary martensite (α'_{prim}). In a second step, the steel is reheated to the so-called partitioning temperature (T_P), where carbon diffuses from the supersaturated α'_{prim} into the untransformed austenite, resulting in its appropriate stabilization upon final cooling to RT [9,10]. In order to ensure the retention of the largest RA fraction, the formation of carbides has to be avoided as much as possible. The addition of Si, Al, or P allows the suppression of cementite precipitation during isothermal holding at T_P, since these elements are considered to be insoluble in cementite [11]. Thus, the cementite growth requires the time-consuming rejection of Si, Al, or P, leading to its postponed precipitation, hence enabling the carbon to partition into austenite. Furthermore, the ε or η transition carbides are known to precipitate already during quenching to T_Q or in the early stages of the partitioning step. However, unlike the influence on cementite formation, the effect of Si, Al, and P on the formation of transition carbides is less clear [12]. From the current research perspective, alloying with these elements does not effectively suppress the precipitation of transition carbides, since they seem to be able to incorporate these elements as solutes [13,14].

The application of the Q&P process to lean medium Mn steels has already been investigated in several studies [15–18]. A beneficial influence of an increased Mn-content on the volume fraction of RA of Q&P steels was confirmed by De Moor [15] and Seo et al. [17]. De Moor et al. [16] has compared the tensile behavior of two 0.3C-3Mn-1.6Si and 0.3C-5Mn-1.6Si steel grades subjected to Q&P heat-treatment and could not state a positive influence of the increased Mn-content on the mechanical properties due to the presence of final martensite in the microstructure of the steel grade containing an elevated Mn level. However, an excellent combination of strength and ductility with UTS × TE exceeding 25,000 MPa was found by Seo et al. [18] for a 0.2C-4.0Mn-1.6Si-1.0Cr Q&P steel.

Since there is still a lack of information regarding the influence of the Mn-content for Q&P steels, this work focuses on the comparison of the microstructure and resulting mechanical properties of three different lean medium Mn Q&P steels containing 0.2 wt.% C, 3.0–4.0 wt.% Mn, and 1.5 wt.% Si. By varying T_Q, the volume fraction of α'_{prim} and thus, RA and α_B were adjusted in order to enable a detailed characterization of the microstructural development and the resulting mechanical properties. Furthermore, since it is not solely the volume fraction of RA, which determines the tensile behavior of TRIP-assisted AHSS, but rather its stabilization against the strain-induced martensite transformation (SIMT), the RA-stability was examined in detail and linked to the stress-strain behavior of the investigated steel grades.

2. Materials and Methods

The chemical compositions of the investigated steel grades with varying Mn-contents are given in Table 1. Three 80 kg ingots were cast under laboratory conditions in a medium frequency furnace, followed by hot rolling to a final thickness of 4 mm. In order to provide cold rollability, the hot rolled sheets (finish rolling temperature = 900 °C) were tempered in a batch-annealing-like furnace for 16 h at 550 °C. Subsequently, the material was cold rolled to a final thickness of 1 mm.

Table 1. Chemical composition of the investigated cold rolled steel grades.

Steel	C	Mn	Si
Fe-C-3.0Mn-Si	0.20	3.06	1.52
Fe-C-3.5Mn-Si	0.20	3.47	1.51
Fe-C-4.0Mn-Si	0.20	3.94	1.50

Specimens of $10 \times 4 \times 1$ mm^3 were heat-treated on a Bähr 805 A/D dilatometer (TA instruments, New Castle, DE, USA) in order to investigate the influence of both, T_Q and Mn-content on the transformation behavior of the cold rolled sheets. Figure 1 shows the applied time-temperature schedules for the Q&P process, adapted to suit an industrially feasible continuous annealing line. After full austenitization for $t_A = 120$ s at $T_A = 850$ °C the samples were quenched to various T_Q in the range of 130–330 °C with a 20 °C step and isothermally held for 10 s (t_Q). Subsequently, partitioning was performed at $T_P = 400$ °C for 300 s, followed by cooling to RT. For tensile testing, strips of $450 \times 20 \times 1$ mm^3 were heat-treated referring to the equal time-temperature regime using a multipurpose annealing simulator (MULTIPAS, voestalpine Stahl GmbH, Linz, Austria), providing electrical resistance heating and gas jet cooling. Three thermocouples were welded onto the strip in order to ensure temperature control.

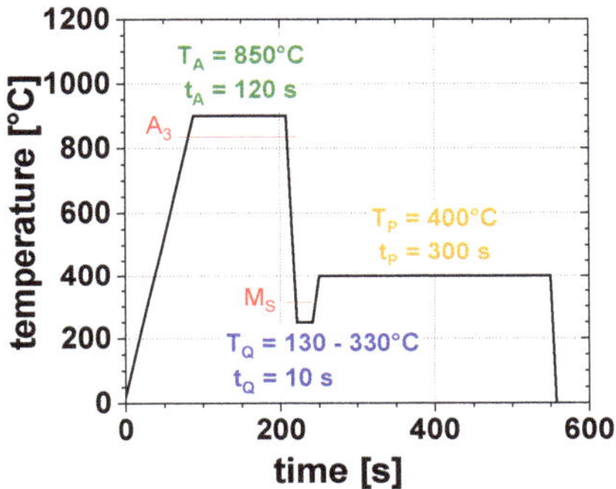

Figure 1. Schematic representation of the applied Quenching and Partitioning (Q&P) heat-treatment cycles for the investigated steels.

Tensile tests were performed according to DIN EN ISO 6892-1, using flat tensile specimens with 25 mm gauge length. For each heat-treatment condition, two samples were tested in a longitudinal direction. The microstructure was characterized by means of LOM using LePera etching. Furthermore, electrochemically polished samples were investigated by means of SEM using a Zeiss SUPRA 35 microscope (Carl Zeiss Microscopy GmbH, Jena, Germany). The volume fracture of RA was determined by means of SMM. Furthermore, interrupted tensile tests were performed at different strain levels to investigate the SIMT and thus, the RA-stability. By means of SMM, the RA-content at gradually increased strains was determined and finally the Ludwigson-Berger relation [19] was used to calculate the k_P-value as an indicator for the RA-stability.

$$\frac{1}{V_\gamma} - \frac{1}{V_{\gamma 0}} = \frac{k_P}{p} * \varepsilon^p \tag{1}$$

Here, V_γ is the volume fraction of RA determined at a specific true strain (ε), $V_{\gamma 0}$ is the initial RA-content before straining and k_P is a factor indicating the RA-stability. p is a strain exponent related to the autocatalytic effect of martensite formation, which can be considered as 1 for TRIP-steels, since this effect can be neglected close to RT [20].

The C-content in RA was determined by the application of the following equation proposed by Dyson and Holmes [21].

$$X_C = \frac{a_\gamma - 3.578 - 0.0056 * X_{Al} - 0.00095 * X_{Mn}}{0.033} \tag{2}$$

Here, X_C is the C-content in RA, a_γ is the austenite lattice parameter in Å, which was measured by X-ray diffraction (XRD) using a PANalytical XPert Pro diffractometer (Malvern Panalytical Ltd, Kassel, Germany) with Co-anode (λ = 0.179 nm, U = 35 kV). X_{Al} and X_{Mn} are the contents of Al and Mn in wt.% in RA, respectively, determined by energy dispersive X-ray spectroscopy (EDX, Carl Zeiss Microscopy GmbH, Jena, Germany).

3. Results

3.1. Transformation Behavior

Figure 2 depicts the dilatometer curves for the investigated steel grades quenched to 3 different T_Q, respectively. It is evident from the graphs that with increasing Mn-content, the M_S-temperature steadily decreased: for the steel grade containing 3.0 wt.% Mn (Figure 2a) the M_S-temperature of 333 °C was determined, for the Fe-C-3.5Mn-Si steel grade (Figure 2b) the M_S-temperature was 315 °C and for the steel grade containing 4.0 wt.% Mn (Figure 2c), an M_S-temperature of 305 °C was observed. Thus, in order to adjust a comparable amount of α'_{prim}, the increase of the Mn-content requires a lower T_Q. As a consequence, the illustrated T_Q were chosen in such a way that regardless of the chemical composition the red curves always correspond to a volume fraction of 50% α'_{prim}, the purple curves represent the samples containing of 75% α'_{prim}, and the blue lines depict the samples where 85% α'_{prim} was adjusted in the microstructure.

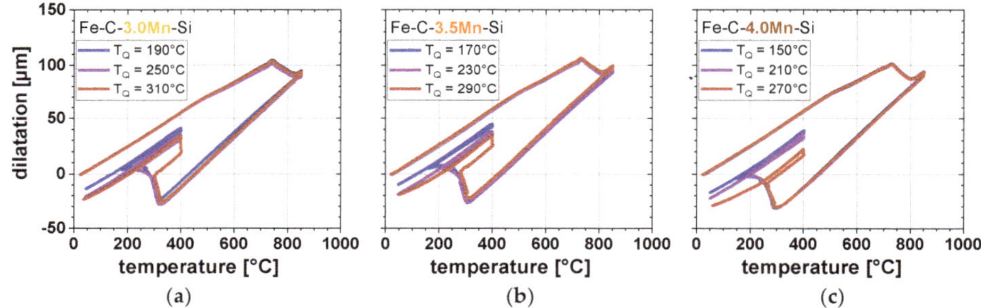

Figure 2. Dilatometric curves for the (**a**) Fe-C-3.0Mn-Si, (**b**) Fe-C-3.5Mn-Si, and (**c**) Fe-C-4.0Mn-Si steels quenched to different T_Q in order to adjust 50 vol.% (red), 75 vol.% (purple) and 85 vol.% (blue) α'_{prim}.

It is obvious from these dilatometric curves that during isothermal holding at T_P γ partially transformed to α_B, accompanied by a linear expansion. Figure 3 gives a detailed perspective on the dilatation during the partitioning step as a function of isothermal holding time. It is evident that with increasing T_Q, and thus decreasing volume fraction of α'_{prim}, a larger amount of α_B was formed. Irrespective of the chemical composition, for the samples with 75% and 85% α'_{prim} (purple and blue lines), rather comparable amount of α_B was formed, whereas for the samples with a matrix consisting of 50% α'_{prim} (red curve) a significant influence of the Mn-content on the formation of α_B could be stated: the higher the Mn-content, the lower the volume fraction of α_B.

In general, for all steels containing 50% α'_{prim}, the formation of α'_{final} was observed during cooling to RT (Figure 2). However, since with increasing Mn-content a decreasing amount of α_B was formed during the partitioning step, the largest volume fraction of α'_{final} was inherently formed for the steel grade containing 4.0 wt.% Mn.

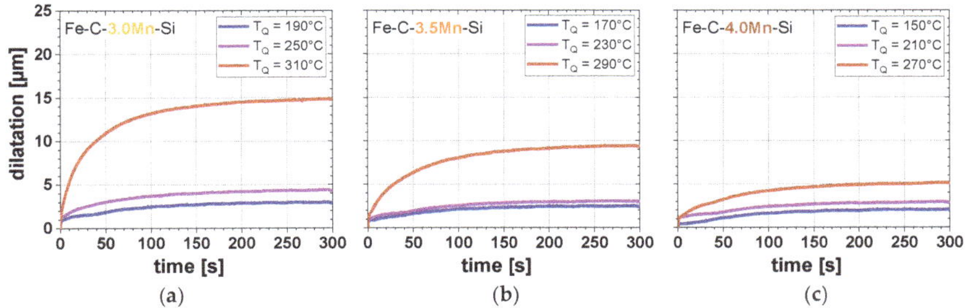

Figure 3. Dilatation due to formation of α_B as a function of isothermal holding time for the (**a**) Fe-C-3.0Mn-Si, (**b**) Fe-C-3.5Mn-Si and (**c**) Fe-C-4.0Mn-Si steels quenched to different T_Q in order to adjust 50 vol.% (red), 75 vol.% (purple), and 85 vol.% (blue) α'_{prim}.

3.2. Microstructure

Figure 4 depicts representative LOM (upper row) and SEM (lower row) images exemplarily shown for the Fe-C-3.5Mn-Si steel grade quenched to $T_Q = 290\,°C$ (Figure 4a,b), $T_Q = 230\,°C$ (Figure 4c,d), and $T_Q = 170\,°C$ (Figure 4e,f). During quenching into the M_S-M_f region α'_{prim} was formed, being subsequently tempered during isothermal holding at T_P. This lath-like carbon-depleted α'' partially included carbide precipitations visible in the SEM images. The microstructural investigation confirmed the results stemming from the analysis of the dilatometric curves, particularly the continuous increase in α'_{prim} and decrease in α_B fraction with declining T_Q. By means of SMM 18.5, 20.7 and 12.5 vol.%, finely distributed RA were measured for the samples quenched to 290, 230, and 170 °C, respectively. Owing to the extremely fine distribution of RA in the tempered-martensitic matrix, its localization is rather difficult using both LOM and SEM. It is obvious from the micrographs that for the sample quenched to 290 °C, a considerable amount of α'_{final} was present in the microstructure due to the insufficient chemical stabilization of austenite. This α'_{final} was characterized in the given microstructure observed by SEM by coarser and surface structured areas. For the steel grades containing 3.0 and 4.0 wt.% Mn, a similar influence of T_Q on the microstructural constituents could be confirmed using LOM and SEM. Regarding the influence of the chemical composition, no significant effect of Mn on the microstructural evolution could be observed at low T_Q, whereas for the higher T_Q, a larger volume fraction of α'_{final} was obtained for the Fe-C-4.0Mn-Si steel, as was also proved by the dilatometric measurements.

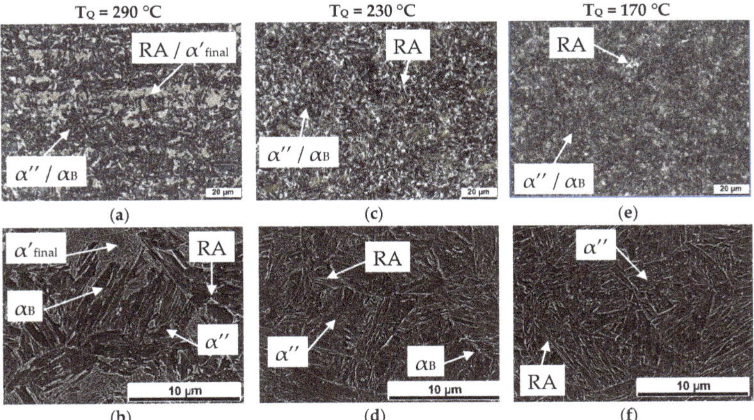

Figure 4. LOM and SEM micrographs of the Fe-C-3.5Mn-Si steel quenched to (**a,b**) $T_Q = 290\,°C$, (**c,d**) $T_Q = 230\,°C$ and (**e,f**) $T_Q = 170\,°C$.

Figure 5 summarizes the results obtained by dilatometry, microstructural investigations, and SMM for the three different steel grades, wherein the individual phase fractions of α'', α_B, RA and α'_{final} are plotted as a function of T_Q. Irrespective of the chemical composition, an increasing T_Q led to a remarkable decrease in the amount of α'', whereas the volume fraction of α_B simultaneously increased. Furthermore, the RA-content rose, until the specific T_Q where α'_{final} was formed as an aftermath of insufficient chemical RA stabilization. This inherently led to a decreasing volume fraction of RA with a further increase of T_Q. It can be regarded from the graphs that with an increasing Mn-content the volume fraction of α_B decreased, leading to a larger amount of RA. For the Fe-C-3.0Mn-Si steel the maximum RA-fraction (RA$_{max}$) of 19.9 vol.% was achieved at $T_Q = 290\ °C$. For the steel grade containing 3.5 wt.% Mn RA$_{max}$ = 22.3 vol.% was measured at $T_Q = 270\ °C$, whereas by further increasing the Mn-content to 4.0 wt.% RA$_{max}$ reached 24.6 vol.% at $T_Q = 230\ °C$. These results indicate, that regardless of the chemical composition, the adjustment of approximately 70 vol.% α'_{prim} was necessary to stabilize the maximum volume fraction of RA. Thus, it can be inferred that the increase of the Mn-content requires the set of lower T_Q. In addition, the value of RA$_{max}$ is positively influenced by an increasing Mn-content as a consequence of the lower fraction of α_B formed during the partitioning step.

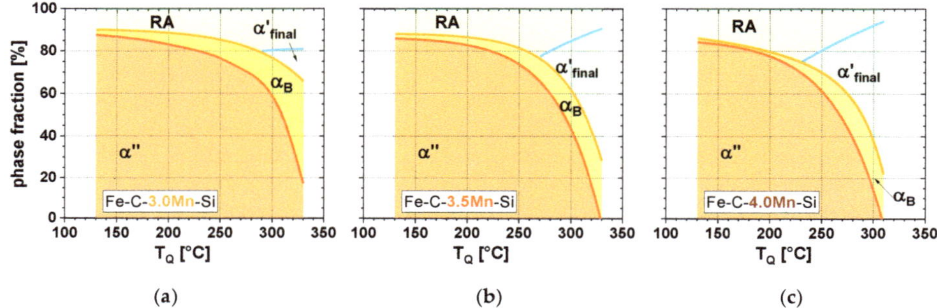

Figure 5. Phase fraction as a function of T_Q for the (**a**) Fe-C-3.0Mn-Si, (**b**) Fe-C-3.5Mn-Si, and (**c**) Fe-C-4.0Mn-Si steels.

3.3. Tensile Testing

Representative engineering stress-strain curves for the investigated steels are presented in Figure 6. For the samples containing approximately 85 vol.% α'_{prim} (blue curves) a comparable stress-strain behavior with very high yield strength (YS), low strain-hardening, and moderate TE was observed. In general, with increasing T_Q a significant decrease in YS was detected for all compositions. When increasing T_Q and thus adjusting 75 vol.% α'_{prim} (purple curves), UTS remained rather constant for the three investigated steels. However, in this case for all compositions an enhanced TE was obtained. When T_Q was further increased, which means that only 50 vol.% α'_{prim} (red curves) was present in the microstructure, a pronounced increase in UTS along with a remarkable decrease in TE was obtained. This was especially in case of the steels containing elevated Mn-contents. The presence of a considerable amount of α'_{final} in the microstructure was particularly responsible for this behavior.

Figures 7 and 8 depict the mechanical properties as a function of T_Q determined by tensile testing. It is clear that independent of the chemical composition first, with increasing T_Q both, YS and UTS decreased until reaching a T_Q of 230 °C (Fe-C-3.0Mn-Si), 210 °C (Fe-C-3.5Mn-Si), and 190 °C (Fe-C-4.0Mn-Si), respectively. This was associated with a decreasing amount of α'_{prim} and thus, increasing volume fraction of RA. Consistent with this, both UE and TE significantly increased, until reaching a sharp maximum, especially in case of the steels containing 3.5 and 4.0 wt.% Mn. Further rise of T_Q led to an increase in UTS, which was more pronounced for the steels with elevated Mn-contents. This was accompanied by a drastic reduction in both UE and TE. In contrast, for the Fe-C-3.0Mn-Si steel in the T_Q-range of 210–290 °C, a rather constant UTS and TE evolution could be obtained. This indicates that with increasing Mn-content, the sensitivity against T_Q fluctuations increased, leading to a narrower process window. Additionally, in case of the 3.5 and 4.0 wt.% Mn steels, at high T_Q, a rise in YS was noticeable, caused by the presence of considerable amounts of α'_{final} formed upon final cooling.

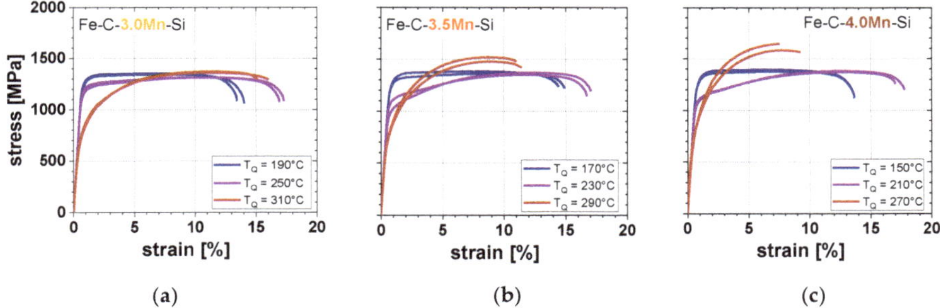

Figure 6. Stress-strain curves for the (**a**) Fe-C-3.0Mn-Si, (**b**) Fe-C-3.5Mn-Si and (**c**) Fe-C-4.0Mn-Si steels quenched to different T_Q in order to adjust 50 vol.% (red), 75 vol.% (purple) and 85 vol.% (blue) α'_{prim}.

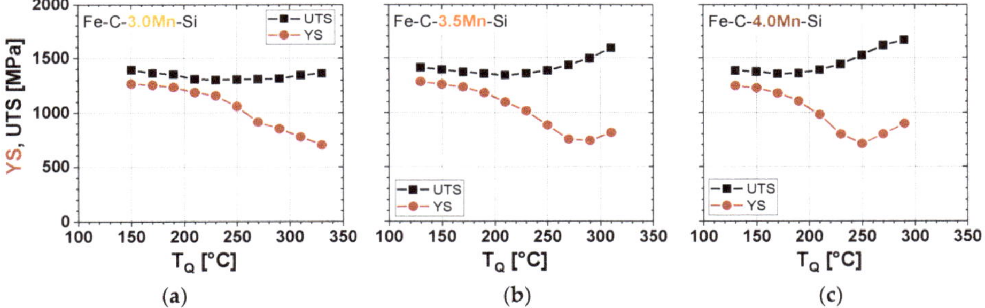

Figure 7. YS and UTS as a function of T_Q for the (**a**) Fe-C-3.0Mn-Si, (**b**) Fe-C-3.5Mn-Si and (**c**) Fe-C-4.0Mn-Si steels.

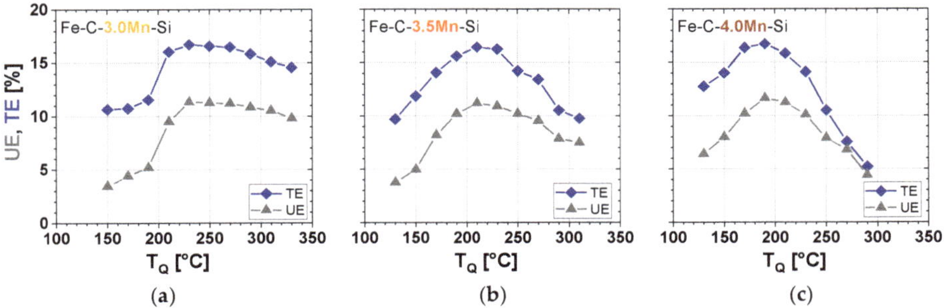

Figure 8. UE and TE as a function of T_Q for the (**a**) Fe-C-3.0Mn-Si, (**b**) Fe-C-3.5Mn-Si, and (**c**) Fe-C-4.0Mn-Si steels

Since the maximum TE was approximately 16.5% for all investigated steels, no evident influence of the Mn-content on the ductility of the investigated steels could be stated. Yet, with increasing Mn-content, a lower T_Q was necessary in order to adjust the maximum TE (T_Q = 230 °C, T_Q = 210 °C, and T_Q = 190 °C for the steels containing 3.0, 3.5, and 4.0 wt.% Mn, respectively). However, a slight influence of Mn on the strength was observed, since with increasing Mn-content UTS rose from 1304 MPa to 1343 MPa and 1360 MPa for the samples achieving the highest TE of 16.5%. Thereby, the best combination of strength and ductility was obtained for the Fe-C-4.0Mn-Si steel with a product of UTS × TE exceeding 22,500 MPa%.

3.4. Retained Austenite Stability

The volume fraction of RA in dependency of true strain is displayed in Figure 9. At a very low T_Q, where approximately 85 vol.% α'_{prim} was adjusted in the microstructure (blue curves), a rather moderate decline in RA-content was observed. This indicates a very stable RA that underwent only minor transformation during straining. The increase of T_Q led to a decline in α'_{prim} fraction to 75 vol.% (purple curves) and therefore to a higher initial volume fraction of RA. This contributed to the pronounced TRIP-effect under these conditions. However, for the samples containing only 50 vol.% α'_{prim} (red curves), a considerable amount of α'_{final} was present in the initial microstructure due to a very low chemical RA-stability. Therefore, in this case, a fast SIMT was observed, especially for the steel grades with elevated Mn-contents.

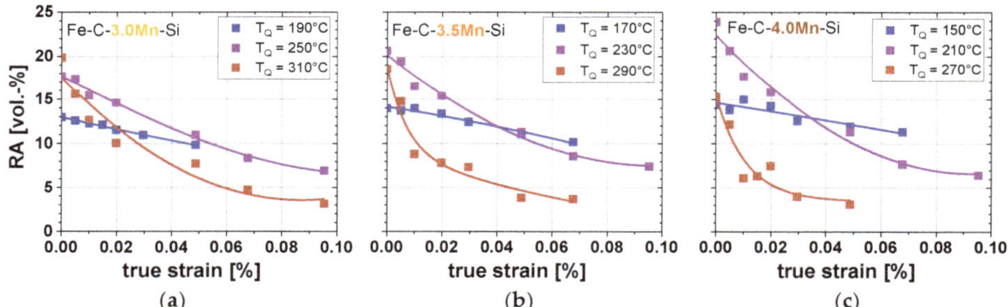

Figure 9. RA-content as a function of true strain obtained after interrupted tensile testing for the (**a**) Fe-C-3.0Mn-Si, (**b**) Fe-C-3.5Mn-Si and (**c**) Fe-C-4.0Mn-Si steels quenched to different T_Q in order to adjust 50 vol.% (red), 75 vol.% (purple) and 85 vol.% (blue) α'_{prim}

In order to enable a better comparison of the RA-stability of the individual steel grades, the volume fraction of γ transformed to α' is plotted as a function of true strain (Figure 10), following a linear relationship according to the Ludwigson-Berger equation (Equation (1)). With rising T_Q increasing k_P-values were observed, indicating a declining RA-stability. For the samples containing 85 vol.% α'_{prim} (blue curves) rather comparable k_P-values between 27 and 48 were determined for the investigated steels, whereas for the samples containing lower α'_{prim} fractions, a substantial influence of the chemical composition was observed. With increasing Mn-content, a significant decrease in RA-stability was found, represented by markedly increasing k_P-values.

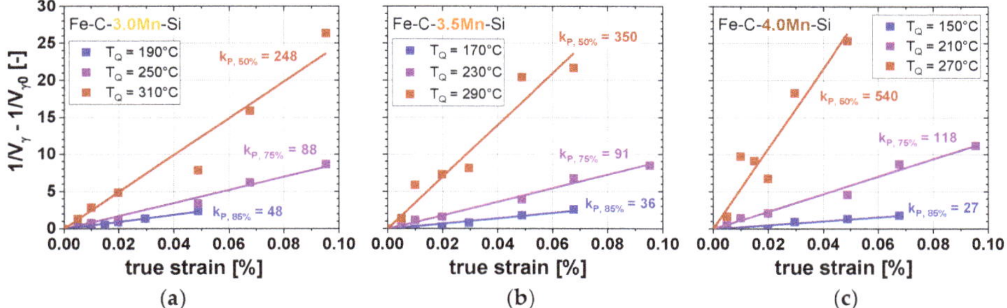

Figure 10. k_P-values indicating the RA-stability of the (**a**) Fe-C-3.0Mn-Si, (**b**) Fe-C-3.5Mn-Si and (**c**) Fe-C-4.0Mn-Si steels quenched to different T_Q in order to adjust 50 vol.% (blue), 75 vol.% (purple), and 85 vol.% (red) α'_{prim}.

In Figure 11a the k_P-values are plotted in dependence of T_Q, summarizing the results displayed in Figure 10. Regardless of the chemical composition, with increasing T_Q continuously rising k_P-values were observed. Since the Mn-content has significant influence on the M_S-temperature, the microstructure of the investigated steel grades consisted of different volume fractions of its individual constituents at comparable T_Q. For this reason, to ensure a better comparability of the steels, Figure 11b presents the k_P-values as a function of α'_{prim}. In general, with decreasing T_Q and hence increasing volume fractions of α'_{prim}, declining k_P-values were observed. This can be attributed to the decreasing fractions of γ_{remain} being present in the microstructure at the onset of the partitioning step. Thus, the C-content in the supersaturated α'_{prim} had to distribute to a lower volume fraction of γ_{remain}, leading to its enhanced stabilization. Furthermore, it is apparent from the graph that for the samples containing volume fractions of α'_{prim} exceeding 70 vol.%, the chemical composition rarely influenced the RA-stability. However, for the samples containing lower α'_{prim} fractions, the increase of the Mn-content led to a pronounced rise in k_P-values related to a substantial decline in RA-stability. The C-content in RA is displayed over the volume fraction of α'_{prim} in Figure 11c. For all steels, with increasing α'_{prim} fraction the C-content in RA rose. For the samples containing at least 70 vol.% α'_{prim} the increase of the Mn-content marginally influenced the C-content in RA. In contrast, at lower α'_{prim} fractions, the influence of the chemical composition was more remarkable, since a sharp decrease in C_γ was observed with increasing Mn-content.

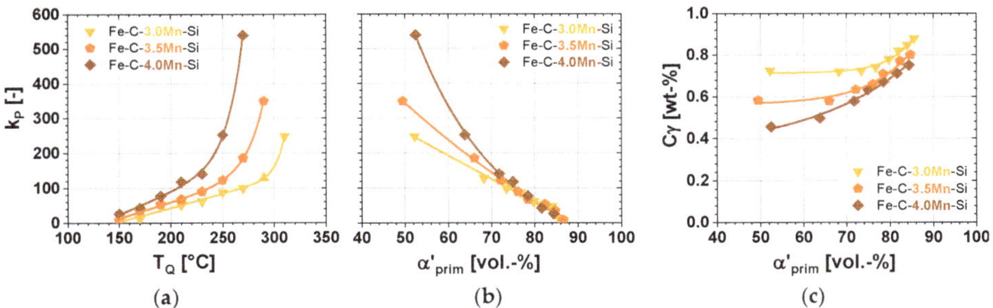

Figure 11. k_P-values indicating the RA-stability as a function of (**a**) T_Q and (**b**) α'_{prim}. (**c**) C-content in RA as a function of α'_{prim} for the investigated steels.

4. Discussion

4.1. Influence of Heat-Treatment Parameter

Regarding the influence of T_Q on the transformation behavior, the results presented in this contribution (Figures 2 and 3) could evidently confirm the findings already known for Q&P steels from the literature [22–25]. With an increasing T_Q, the volume fraction of α'_{prim} gradually decreased as a result of the reduced driving force for the $\gamma \rightarrow \alpha'$ transformation, as described by Koistinen and Marburger [26]. Coinciding with the decreasing amount of α'_{prim}, a larger volume fraction of γ_{remain} was present in the microstructure at the onset of isothermal holding at T_P. Hence, with a rising T_Q, a larger amount of γ_{remain} transformed to α_B. Furthermore, by increasing T_Q, a rising volume fraction of RA (Figure 5) was stabilized until exceeding a critical T_Q, leading to the formation of α'_{final} owing to the insufficient stabilization of γ_{remain}. Thus, in accordance with Speer et al. [27], a triangular shape of the RA-fraction as a function of T_Q could be observed (Figure 5).

The considerable influence of T_Q on the microstructural constituents obviously affected the mechanical properties of the investigated steels. With increasing T_Q, a decline in both UTS and YS accompanied by a rise in TE could be observed (Figures 6–8), which is consistent with the findings of De Moor et al. [12]. This behavior is linked to the decreasing volume fractions of α'_{prim} and in turn to rising RA-contents. Therefore, it is inferred that at very low T_Q, RA was hyper-stable, i.e., it underwent almost no transformation during deformation (Figures 9–11). With increasing T_Q the chemical RA-stability decreased (Figure 11c), resulting in a pronounced TRIP-effect, which contributed to the enhanced combination of strength and ductility. Independent of the chemical composition, the best combination of UTS and TE was observed 40 °C below a T_Q amount where the maximum RA-content could be stabilized at RT (T_Q = 230 °C, 210 °C, and 190 °C for the steels containing 3.0, 3.5, and 4.0 wt.% Mn, respectively). At a higher T_Q than the optimal one, a significant increase in UTS associated with a sharp decline in TE was observed, resulting in a remarkable deterioration of UTS × TE. This was linked to the presence of α'_{prim} in the initial microstructure related to low chemical RA-stability. Therefore, in this case, RA could not contribute to the enhanced TRIP-effect, which is in correlation with results reported in literature [28–30].

4.2. Influence of Mn-Content

Concerning the main topic of this contribution, the influence of the Mn-content on microstructure and resulting mechanical properties, interesting findings could be observed. First, the effect of an increasing Mn-content on the decrease in M_S-temperature reported in literature [31,32] was confirmed by means of dilatometry. Therefore, for the steels containing elevated Mn-content, lower T_Q were necessary in order to adjust comparable amounts of α'_{prim} (Figure 2).

In the case of the samples containing at least 80 vol.% α'_{prim}, only very low volume fractions of α_B were formed during isothermal holding at T_P. The increase of the Mn-content from 3.0 to 3.5 led to a slight decrease in volume fraction of α_B, and thus to a minor increase in RA-content (Figure 12a). These marginal differences in microstructure were directly reflected in the mechanical properties of the Fe-C-3.0Mn-Si and Fe-C-3.5Mn-Si steels, since the slightly increased volume fractions of RA contributed to a more pronounced TRIP-effect. As a consequence, a slight rise in UTS × TE was observed by increasing the Mn-content from 3.0 to 3.5 wt.% (Figure 12b) for the samples containing at least 80 vol.% α'_{prim}. Regarding the RA-stability, almost no differences in k_P-values were measured for these samples (Figures 10 and 11b). In counterpart, at comparable volume fractions of α'_{prim}, the C-content in RA shown in Figure 11c) was slightly lower for the steel containing 3.5 wt.% Mn, compared to the Fe-C-3.0Mn-Si steel grade. This could be attributed to the marginally larger RA-contents, leading to a lower C-enrichment during partitioning. In this context, higher products of RA-fraction and C-content in RA were achieved by an increase of the Mn-content from 3.0 to 3.5 wt.% (Figure 12c). On the contrary, the further increase of the Mn-content from 3.5 to 4.0 wt.% barely influenced the volume fraction of RA (Figure 12a). Therefore, the mechanical properties (Figure 12b) as well as the product of C_γ and RA (Figure 12c) remained almost unchanged.

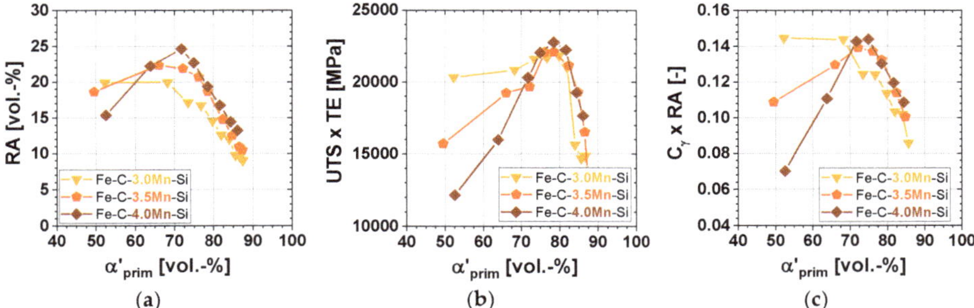

Figure 12. (**a**) RA-content, (**b**) UTS × TE and (**c**) C_γ × RA as a function of α'_{prim} for the investigated steels.

In general, in case of the samples containing at least 80 vol.% α'_{prim}, the C-diffusion from the supersaturated α'_{prim} to the γ_{remain} was the predominant mechanism for RA-stabilization. In contrast, the formation of α_B barely contributed to the C-enrichment of γ_{remain} due to its relatively low fraction. Furthermore, apart from the C-content in RA, its chemical stability is improved by an increased Mn-content in RA, as well. Since no evident Mn-parititoning during isothermal holding at T_P was observed by means of EDX, the higher Mn-content in the bulk composition counterbalanced the lower C_γ and inherently enhanced the chemical RA-stability for the steels containing 3.5 and 4.0 wt.% Mn. Nevertheless, it has to be considered that Mn influences the carbide formation in steels, since Mn is soluble in cementite [33]. This could lead to the precipitation of larger amounts of carbides during heat-treatment with increasing Mn-content, acting as carbon sinks and thus reducing the capacity for partitioning of C into γ. Hence, further research efforts are necessary to investigate the influence of Mn on the carbide precipitation and as a consequence C partitioning from the supersaturated α'_{prim} to γ_{remain} in the case of the present steels.

For the samples quenched to higher T_Q, in order to adjust 70–80 vol.% α'_{prim}, the influence of Mn on the phase transformation was more apparent compared to those containing higher fractions of α'_{prim}. This was due to the larger volume fractions of γ_{remain} being present at the onset of isothermal holding at T_P, influencing both the partitioning process and the formation of α_B. (Figures 2 and 3). According to the T_0-concept, a thermodynamic limit exists for the $\gamma \rightarrow \alpha$ transformation [34]. As γ_{remain} is enriched in C during the α_B formation, a diffusion-free transformation is thermodynamically impossible, as soon as the carbon content reaches a critical value. Since the difference in Gibb's free energy (ΔG_m) between γ and α is the driving force for the $\gamma \rightarrow \alpha$ transformation, the bainitic reaction stops if ΔG_m reaches 0 due to the C-enrichment in γ. This point is determined by the T_0-temperature. It is well known that Mn shifts the T_0-line to lower C-contents [35], allowing the formation of lower amounts of α_B in case of an increasing Mn-content owing to the above-mentioned thermodynamic limit. However, this effect is negligible compared to the general retarding effect of Mn on the bainitic transformation kinetics [36,37]. Mn enriches at the former austenitic grain boundaries, hindering the ferritic nucleation due to the local decrease in A_{e3}. In addition, Mn reduces the diffusion rate of C in γ, which has a clear retardation effect on the formation of α_B [38]. Therefore, in case of the samples containing 70–80 vol.% α'_{prim} the increase of the Mn-content led to lower fractions of α_B, and in turn to a significant rise in RA-content (Figure 12a). This increase in volume fraction of RA was accompanied by its decreasing chemical stabilization as proven by interrupted tensile tests (Figures 10 and 11), since the overall C-content had to partition to a larger volume fraction of γ_{remain}. Irrespective of the steel composition, the remarkably lower RA-stability compared to the samples quenched to lower T_Q contributed to the pronounced TRIP-effect, leading to enhanced strain hardening and thus improved combinations of UTS and TE (Figure 12b). Regarding the influence of the Mn-content, only minor differences in terms of mechanical properties could be found for these samples. This was linked to rather small deviations of $C_\gamma \times$ RA-content (Figure 12c), especially in the case of the samples containing 3.5 and 4.0 wt.% Mn.

However, for the samples containing even lower volume fractions of α'_{prim} (<70 vol.%), a substantial influence of the chemical composition on the phase transformation behavior and resulting structure-properties relation was found. In case of the Fe-C-3.0Mn-Si steel, the formation of up to 25 vol.% α_B (Figure 5) led to a moderate increase in RA-content along with decreasing α'_{prim} fractions. Since for the samples containing elevated Mn-contents the bainitic transformation was delayed, significant larger volume fractions of γ_{remain} transformed to α'_{final} during final cooling to RT, resulting in a sharp decrease in RA-contents with declining amounts of α'_{prim} (Figure 12a). In case of the Fe-C-3.0Mn-Si steel, this resulted in rather constant and high values of $C_\gamma \times$ RA (Figure 12c), whereas with increasing Mn-contents a drastic drop was observed, caused by the considerably declining RA-contents. Obviously, this was reflected in the stress-strain behavior of the investigated steels. For the composition containing 3.0 wt.% Mn, the mechanical properties remained almost unchanged by the increase of T_Q, resulting in almost consistent values of UTS $\times$ TE (Figure 12b). In contrast, for the Fe-C-3.5Mn-Si and Fe-C-4.0Mn-Si steels containing less than 70 vol.% α'_{prim}, the presence of α'_{final} related to very low RA-stabilities resulted in a significant increase in UTS, accompanied by a radical loss in TE. These results are coherent with those published by De Moor et al. [15] who reported a high sensitivity against T_Q-fluctuations in case of Q&P steels containing elevated Mn-contents. On that account, the formation of increased α_B fractions in case of the Fe-C-3.0Mn-Si steel contributed to the wider process window in terms of constant mechanical properties, compared to the steels containing 3.5 and 4.0 wt.% Mn.

5. Conclusions

This contribution focused on the study of the influence of the Mn-content on the microstructural evolution and mechanical properties of lean medium Mn Q&P steels containing 0.2% C, 1.5% Si, and 3.0–4.0 wt.% Mn, with a special emphasis on the RA-stability. The findings of the present paper are as follows:

- Regardless of the chemical composition, by increasing T_Q the volume fraction of α'_{prim} steadily decreased, accompanied by a rising amount of α_B and RA. The exceedance of a critical T_Q, depending on the Mn-content, resulted in an insufficient chemical stabilization of RA, triggering the formation of α'_{final} during final cooling to RT.

- A significant influence of the Mn-content on the phase transformation behavior could be observed, particularly with increasing T_Q and thus decreasing α'_{prim} fraction. The addition of enhanced Mn-contents led to an appreciable delay in $\gamma \rightarrow \alpha_B$ transformation during the partitioning step. Thus, on the one hand, larger volume fractions of RA could be stabilized with increasing Mn-content. On the other hand, the increase of the Mn-content adversely affected the RA-stability due to the declining C-content in RA, which was only partially counterbalanced by the enhanced Mn-content in RA.

- The mechanical properties achieved by the Q&P process were pronouncedly determined by both, volume fraction and stability of RA. With increasing Mn-content, a remarkably stronger sensitivity against T_Q-fluctuations in terms of RA-content and its stability was observed. As a result, the increase of the Mn-content resulted in a narrower process window with regard to the robustness of mechanical properties.

- For all investigated steels, the best combination of UTS and TE was observed for microstructures containing 75–80 vol.% α'_{prim}. For this reason, a T_Q 40 °C below the maximum RA-content had to be set in order to obtain the optimum mechanical properties. By increasing the Mn-content, the maximum value of UTS $\times$ TE could exceed 22,500 MPa%, since the larger volume fraction of RA by approximately 5% contributed to an enhanced TRIP-effect.

Author Contributions: Conceptualization, R.S., D.K., and C.S.; methodology, S.K.; software, S.K.; validation, D.K. and R.S.; formal analysis, S.K.; investigation, S.K.; resources, D.K., R.S., and S.K.; data curation, S.K., R.S., D.K., C.B., and C.S.; writing-original draft preparation, S.K.; writing-review and editing, R.S., D.K., C.B., and C.S.; visualization, S.K.; supervision, R.S., D.K., C.B., and C.S.; project administration, D.K., R.S., and C.S.; funding acquisition, D.K., R.S.

Funding: This research was funded by the Austrian Research Promotion Agency (FFG), grant number 860188, "Upscaling of medium Mn-TRIP steels".

Conflicts of Interest: The authors declare no conflict of interest. The funders had no role in the design of the study; in the collection, analyses, or interpretation of data; in the writing of the manuscript, or in the decision to publish the results.

Abbreviations

α	ferrite
α'	martensite
α'_{final}	final martensite
α'_{prim}	primary martensite
α''	tempered martensite
α_B	bainitic ferrite
a_γ	austenite lattice parameter
C_γ	carbon content in retained austenite
γ_{remain}	remaining austenite
ΔG_m	difference in Gibb's free energy
EDX	energy dispersive X-ray spectroscopy
k_P	factor indicating the RA-stability
LOM	light optical microscopy
MULTIPAS	multipurpose annealing simulator
p	strain exponent related to the autocatalytic effect
Q&P	quenching & partitioning
RA	retained austenite
RA_{max}	maximum retained austenite
RT	room temperature
SEM	scanning electron microscopy
SIMT	strain induced martensitic transformation
TE	total elongation
T_P	partitioning temperature
T_Q	quenching temperature
t_Q	quenching time
TRIP	transformation induced plasticity
UTS	ultimate tensile strength
$V_{\gamma 0}$	initial volume fraction of retained austenite
X_{Al}	aluminum content in retained austenite
X_C	carbon content in retained austenite
X_{Mn}	manganese content in retained austenite
XRD	X-ray diffraction
YS	yield strength

References

1. Fonstein, N. *Advanced High Strength Sheet Steels*, 1st ed.; Springer International Publishing: Cham, Switzerland, 2015; pp. 5–7.
2. Kwon, O.; Lee, K.; Kim, G.; Chin, K. New trends in advanced high strength steel - Developments for automotive application. *Mater. Sci. Forum* **2010**, *638–642*, 136–141. [CrossRef]
3. Steineder, K.; Krizan, D.; Schneider, R.; Béal, C.; Sommitsch, C. On the microstructural characteristics influencing the yielding behavior of ultra-fine grained medium-Mn steels. *Acta Mater.* **2017**, *139*, 39–50. [CrossRef]

4. Matlock, D.; Speer, J.; De Moor, E.; Gibbs, P. Recent developments in advanced high strength steels for automotive applications: An overview. *JESTECH* **2012**, *15*, 1–12.
5. De Cooman, B.C.; Speer, J. Quench and partitioning steel: A new AHSS concept for automotive anti-intrusion applications. *Steel Res. Int.* **2006**, *77*, 634–640. [CrossRef]
6. Steineder, K.; Krizan, D.; Schneider, R.; Béal, C.; Sommitsch, C. On the damage behavior of a 0.1C6Mn Medium-Mn Steel. *Steel Res. Int.* **2017**, *89*, 1700378. [CrossRef]
7. Steineder, K.; Schneider, R.; Krizan, D.; Béal, C.; Sommitsch, C. Comparative investigation of phase transformation behavior as a function of annealing temperature and cooling rate of two medium-Mn steels. *Steel Res. Int.* **2015**, *85*, 1–8. [CrossRef]
8. Arlazarov, A.; Gouné, M.; Bouaziz, O.; Hazotte, A.; Petitgand, G.; Berger, P. Evolution of microstructure and mechanical properties of medium Mn steels during double annealing. *Mater. Sci. Eng. A* **2012**, *542*, 31–39. [CrossRef]
9. Speer, J.; Matlock, D.; De Cooman, B.C.; Schroth, J. Carbon partitioning into austenite after martensite transformation. *Acta Mater.* **2003**, *51*, 2611–2622. [CrossRef]
10. Speer, J.; Aussuncao, F.; Matlock, D.; Edmonds, D. The quenching and partitioning process: Background and recent progress. *Mater. Res.* **2005**, *51*, 2611–2622. [CrossRef]
11. Owen, W. Effect of silicon on the kinetics of tempering. *Trans. ASM* **1954**, *46*, 812–829.
12. De Moor, E.; Lacroix, S.; Clarke, A.; Penning, J.; Speer, J. Effect of Retained Austenite Stabilized via Quench and Partitioning on the Strain Hardening of Martensitic Steels. *Metall. Mater. Trans. A* **2008**, *39A*, 2586–2595. [CrossRef]
13. Speich, G.; Leslie, W. Tempering of steel. *Metall. Trans.* **1972**, *3*, 1043–1054. [CrossRef]
14. Krauss, G. Tempering and structural change in ferrous martensitic structures. In *Phase Transformations in Ferrous Alloys: Proceedings of an International Conference, Proceedings of the International Conference on Phase Transformations in Ferrous Alloys, TMS-AIME, Warrendale, PA, USA, 1984*; Marder, A.R., Goldstein, J.I., Eds.; AIME: Englewood, CO, USA, 1984; pp. 101–123.
15. De Moor, E.; Speer, J.; Matlock, D.; Kwak, J.; Lee, S.-B. Effect of carbon and manganese on the quenching and partitioning response of CMnSi steels. *ISIJ Int.* **2011**, *51*, 137–144. [CrossRef]
16. De Moor, E.; Speer, J.; Matlock, D.; Kwak, J.-H.; Lee, S.-B. Quenching and partitioning of CMnSi steels containing elevated manganese levels. *Steel Res. Int.* **2012**, *83*, 322–327. [CrossRef]
17. Seo, E.-J.; Cho, L.; De Cooman, B.C. Application of quenching and partitioning processing to medium Mn steel. *Metall. Mater. Trans. A* **2015**, *46*, 27–31. [CrossRef]
18. Seo, E.-J.; Cho, L.; De Cooman, B.C. Kinetics of the partitioning of carbon and substitutional alloying elements during quenching and partitioning (Q&P) processing of medium Mn steel. *Acta Mater.* **2016**, *107*, 354–365.
19. Ludwigson, D.; Berger, J. Plastic behaviour of metastable austenitic stainless steels. *J. Iron Steel Inst.* **1969**, *207*, 63–69.
20. Matsumura, O.; Sakuma, Y.; Takechi, H. TRIP and its kinetic aspects in austempered 0.4C-1.5Si-0.8Mn steel. *Scr. Metall.* **1987**, *21*, 1301–1306. [CrossRef]
21. Dyson, D.; Holmes, B. Effect of alloying additions on the lattice parameter austenite. *J. Iron Steel Inst.* **1970**, *208*, 469–474.
22. Kaar, S.; Schneider, R.; Krizan, D.; Béal, C.; Sommitsch, C. Influence of the quenching and partitioning process on the transformation kinetics and hardness in a lean medium manganese TRIP steel. *Metals* **2019**, *9*, 353. [CrossRef]
23. Kaar, S.; Schneider, R.; Krizan, D.; Béal, C.; Sommitsch, C. Influence of the phase transformation behaviour on the microstructure and mechanical properties of a 4.5 wt.-% Mn Q&P steel. *HTM J. Heat Treatm. Mat.* **2010**, *74*, 70–83.
24. Santofimia, M.; Zhao, L.; Petrov, R.; Kwakernaak, C.; Sloof, W.; Sietsma, J. Microstructural development during the quenching and partitioning process in a newly designed low-carbon steel. *Acta Mater.* **2011**, *59*, 6059–6068. [CrossRef]
25. HajyAkbary, F.; Santofimia, M.; Sietsma, J. Optimizing mechanical properties of a 0.3C-1.5Si-3.5Mn quenched and partitioned steel. *Adv. Mater. Res.* **2014**, *829*, 100–104. [CrossRef]
26. Koistinen, D.; Marburger, R. A general equation prescribing the extent of the austenite-martensite transformation in pure iron-carbon alloys and plain carbon steels. *Acta Metall.* **1959**, *7*, 59–60. [CrossRef]

27. Speer, J.; Streicher, A.; Matlock, D.; Rizzo, F. Quenching and partitioning: A fundamentally new process to create high strength TRIP sheet microstructures. In Proceedings of the Austenite Formation and Decomposition MS&T, Chicago, IL, USA, 9–12 November 2003; pp. 505–522.
28. De Knijf, D.; Petrov, R.; Föjer, C.; Kestens, L. Effect of fresh martensite on the stability of retained austenite in quenching and partitioning steel. *Mater. Sci. Eng. A* **2014**, *615*, 107–115. [CrossRef]
29. Steineder, K.; Krizan, D.; Schneider, R.; Béal, C.; Sommitsch, C. The effects of intercritical annealing temperature and initial microstructure on the stability of retained austenite in a 0.1C-6Mn steel. *Mater. Sci. Forum* **2016**, *879*, 1847–1852. [CrossRef]
30. Seo, E.; Cho, L.; Estrin, Y.; De Cooman, B.C. Microstructure-mechanical properties relationships for quenching and partitioning (Q&P) processed steel. *Acta Mater.* **2016**, *113*, 124–139.
31. Mahieu, J.; Maki, J.; De Cooman, B.C.; Claessens, S. Phase transformation and mechanical properties of Si-free CMnAl transformation-induced plasticity-aided steel. *Metall. Mater. Trans. A* **2002**, *33*, 2573–2580. [CrossRef]
32. Schneider, R.; Steineder, K.; Watanebe, A.; Okumiya, M.; Krizan, D.; Sommitsch, C. Determination of a new empirical M_S-formula suitable for medium-Mn-steels. In Proceedings of the 24th IFHTSE Congress 2017—European Conference on Heat Treatment and Surface Engineering, Nice, France, 26–29 June 2017; pp. 1–9.
33. Satzinger, K. *Einfluss von Chrom und Mangan auf die Bainitbildung in Dualphasenstählen*; Diplomarbeit, Montanuniversität Leoben: Leoben, Austria, 2008.
34. Röthler, B. Möglichkeiten zur Beeinflussung der Mechanischen Eigenschaften von Kaltgewalzten TRIP-Stählen. Ph.D. Thesis, Technische Universität München, München, Germany, 2005.
35. Paul, S. Entwicklung Neuer Legierungskonzepte für Höchstfeste TRIP-Stähle mit Nicht Ferritischer Matrix und Reduziertem Siliziumgehalt. Ph.D. Thesis, Technische Universität München, München, Germany, 2012.
36. Eggbauer, G. Charakterisierung bainitischer Gefügezustände für Gesenkschmiedeteile. *BHM* **2014**, *159*, 5, 194–200.
37. Bleck, W.; Moeller, E. *Handbuch Stahl: Auswahl, Verarbeitung, Anwendung*, 1st ed.; Carl Hanser Verlag: Rastatt, Germany, 2017; pp. 292–293.
38. De Moor, E.; Matlock, D.; Speer, J.; Merwin, M. Austenite stabilization through manganese enrichment. *Scr. Mat.* **2011**, *64*, 185–188. [CrossRef]

Article

The Impact of Strain Heterogeneity and Transformation of Metastable Austenite on Springback Behavior in Quenching and Partitioning Steel

Yonggang Yang [1,*], Zhenli Mi [1,*], Siyang Liu [2], Hui Li [3], Jun Li [4] and Haitao Jiang [1]

[1] Institute of Engineering Technology, University of Science and Technology Beijing, Beijing 100083, China; Jianght_ustb@163.com

[2] School of Advanced Engineering, University of Science and Technology Beijing, Beijing 100083, China; sepxiu2011@163.com

[3] College of Engineering, Yantai Nanshan University, Yantai 265700, China; lihui9@nanshan.edu.cn

[4] Materials Department of Automotive Engineering Research Institute, Chery Automobile Co., Ltd., Wuhu 241009, China; lijun6@mychery.com

* Correspondence: yeungyg@163.com (Y.Y.); mizl@nercar.ustb.edu.cn (Z.M.); Tel.: +86-10-6233-6603 (Z.M.); Fax: +86-10-6233-2947 (Z.M.)

Received: 3 May 2018; Accepted: 5 June 2018; Published: 7 June 2018

Abstract: Multiple strengthening methods, such as high dislocation density, high twin density, small grain size, and metastable austenite phase can give high strength to ultra-high strength steels (UHSSs). However, the high strength of UHSSs often results in a greater tendency for springback when applied in manufacturing vehicle components. In the present study, two types of UHSSs, dual-phase (DP) steel and quenching and partitioning (QP) steel are investigated to study the springback behavior during the bending process. Results indicated that both the strain heterogeneity and the transformation of retained austenite impacted the springback behavior. The springback angle of the DP steel increased with the increase in bending angle, which was caused by the increasing degree of strain heterogeneity. However, the springback angle of the QP steel decreased to a 14.75° when QP specimens were strained at a 104° bending angle due to the inhibiting effect of the phase transformation. This indicated that there was preferential phase transformation in the thickness direction in the retained austenite of the outer and inner zones. The phase transformation caused low strain heterogeneity, which resulted in a lower tendency for springback. The results suggested that QP steel could possess lower springback at a proper bending angle.

Keywords: quenching and partitioning steel; springback behavior; strain heterogeneity; metastable austenite

---

1. Introduction

Recently, ultra-high strength steels (UHSSs) have attracted much attention from researchers and automobile manufacturers because of their remarkable properties of high strength and ductility. Compared to conventional steel, the higher strength of UHSSs ensures there is greater potential for weight reduction [1–3]. Particularly, in order to accomplish a better balance of strength and ductility, Speer et al. [4] proposed the quenching and partitioning (QP) process to obtain stabilized austenite in a martensitic microstructure. As a result of the existence of transformation-induced plasticity (TRIP) effect, the newly developed QP steel also possesses high strength and great ductility with few alloy additions [5,6]. However, the high strength achieved by high strain hardening in UHSSs often results in a greater tendency for springback when applied in manufacturing vehicle components for the automotive industry using stamping technologies [7].

For accurate springback prediction, several material models are modified based on material properties, such as the Bauschinger effect [8], transient behavior [9], nonlinear elastic behavior [10], and modulus change behavior [11]. Based on the material properties, Eggertsen and Mattiasson [12] modified the yield functions and the hardening laws and concluded that the yield functions and the hardening laws have a greater impact on the predicted springback. Yang et al. [13] also reported that taking into account the change of Young's modulus and obtaining a precise hardening function facilitate the accuracy of springback prediction. Furthermore, Wang et al. [14] simulated the loading process more accurately using a non-saturating kinematic Swift model and adopted inertia relief as a new control approach in springback calculation for the unloading process. However, there are still some problems in springback prediction of TRIP-assisted steel. Lee et al. [15,16] combined the homogeneous anisotropic hardening model and quasi-plastic-elastic strain formulations to describe the elasticity and plasticity behavior of DP steel and TRIP steel. The combined model characterizes flow curves of DP steel and predicts the springback very well. However, springback prediction is not good for TRIP steel, mainly due to its deformation-induced martensitic transformation-dependent flow behavior. Hence, the transformation of metastable austenite is considered to be a crucial essential in predicting the springback of TRIP-assisted UHSSs. The transformation of metastable austenite may cause the springback prediction of TRIP-assisted UHSSs to become more difficult. However, less attention has been focused on the relation between the transformation of metastable austenite and the springback behavior of TRIP-assisted steel so far [17,18].

The aim of this paper is to investigate the springback behavior of quenching and partitioning steel during the bending process. Moreover, the relationship between the springback behavior and the strain heterogeneity was analyzed. Additionally, the impact of transformation of metastable austenite on the springback was further investigated using dual-phase steel without transformation of metastable austenite as a reference.

2. Experimental Procedure

The study focused on 1300-MPa-grade cold-rolled QP steel and 1200-MPa-grade DP steel. The cold-rolled QP and dual phase (DP) steels, with a 1.2-mm thickness, were provided by Hansteel. The compositions of QP and DP steel are listed in Table 1. The bending test samples were cut from the QP and DP steel along the rolling direction with length and width of 66 mm and 20 mm, respectively, in accordance with the VDA238-100 standard. Bending tests were conducted with a deformation rate of 2 mm/min on a CMT 5105 universal testing machine (SANS Testing Machine Co., Shenzhen, China), as displayed in Figure 1a–c. During the bending process, DP and QP specimens were bent to different press depths. Pictures were taken when the specimen was under bending and after springback. Based on the photos of specimens before and after springback, the springback angles of DP and QP steel were calculated. Three specimens at every bending angle were calculated. The accuracy of measurement was ensured based on the repeated specimens. Furthermore, the bending angles were calculated based on Equations (1)–(4) [19].

$$g = (R + \frac{L}{2})^2 + (R + a - S)^2 \tag{1}$$

$$h = 2(R + a)^2 \cdot \left[-(R + \frac{L}{2}) \right] + 2(R + \frac{L}{2})^3 - 2(R + a - S)^2 \cdot \left[-(R + \frac{L}{2}) \right] \tag{2}$$

$$i = (R + a)^4 - 2(R + a)^2 \cdot (R + \frac{L}{2})^2 - (R + a - S)^2 \cdot (R + a)^2 + (R + a - S)^2 \cdot (R + \frac{L}{2})^2 + (R + \frac{L}{2})^4 \tag{3}$$

$$\alpha = 2 \cdot \left\{ -\arctan \left[\frac{\sqrt{(R + a)^2 - (\frac{-\sqrt{h^2 - 4g \cdot i} - h}{2g} + R + \frac{L}{2})^2} - (R + a - S)}{\frac{-h - \sqrt{h^2 - 4g \cdot i}}{2g}} \right] \cdot \frac{180}{\pi} \right\} \tag{4}$$

where R and L represent the roll radius and the distance of two rolls, respectively; S and a are the stroke depth and the specimen thickness, respectively; α is the bending angle of the specimens; and g, h, and i are the intermediate variables of the calculation process.

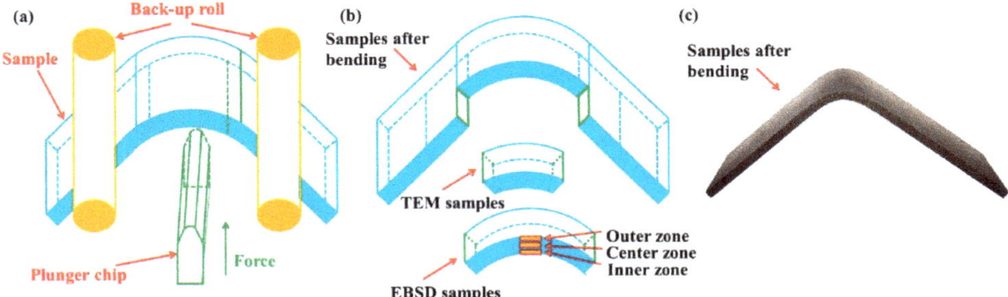

Figure 1. Diagrams describing the bending process and locations of electron backscattered diffraction (EBSD) and transmission electron microscopy (TEM) specimens. (**a**) Schematic diagram of the bending process; (**b**) Schematic diagram describing locations of EBSD and TEM specimens. The yellow rectangles represent the outer zone, center zone, and inner zone in the thickness direction; (**c**) Diagram of the bending samples.

Table 1. Main chemical compositions and mechanical properties of materials used in experiments. DP: dual phase, QP: quenching and partitioning.

Sample	C	Mn	Si	P	S	Nb	Fe	Yield Strength	Tensile Strength
				wt %					MPa
DP	0.15	1.83	0.25	0.01	0.001	0.028	Balance	961	1192
QP	0.23	2.31	1.60	0.018	0.003	0.031	Balance	1209	1364

Transmission electron microscopy (TEM, F20 FEI, Hillsboro, OR, USA) and electron backscattered diffraction (EBSD, Nordlys-II & Channel 5.0, Oxford, UK) were used to characterize the microstructure. TEM and EBSD specimens were cut from the bending specimens as shown in Figure 1b. After mechanical grinding, EBSD samples were polished and electro-polished at room temperature using a solution containing 95 vol % C_2H_5OH and 5 vol % $HClO_4$. The TEM samples were mechanically ground to a thickness of 50 μm and then electro-polished with a twin-jet electropolisher (Struers TenuPol-5, Copenhagen, Denmark) at $-20\ ^\circ C$ using a solution containing 85 vol % C_2H_5OH and 15 vol % $HClO_4$. The retained austenite (RA) volume fraction of QP steel was measured using X-ray diffraction (XRD, STOE, Darmstadt, Germany) and calculated based on the integrated intensity of $(200)\alpha$, and $(211)\alpha$ ferrite/martensite peaks and $(200)_\gamma$, $(220)_\gamma$, and $(311)_\gamma$ austenite peaks.

3. Results and Discussion

3.1. Initial Microstructures

The microstructure of DP and QP steel prior to deformation is shown in Figure 2. As depicted in Figure 2a,b, the DP steel sheets had lath-like martensite and island-like ferrite, distributed homogeneously across the material. The microstructure of the QP steel contained martensite and retained austenite (Figure 2c,d). The retained austenite was highlighted by red arrows in Figure 2c. The retained austenite volume fraction in QP steel was 15.9%, which was evaluated using XRD and EBSD analysis. The difference between the results of both RA volume fraction measurement techniques was small (i.e., maximum of 1.4%).

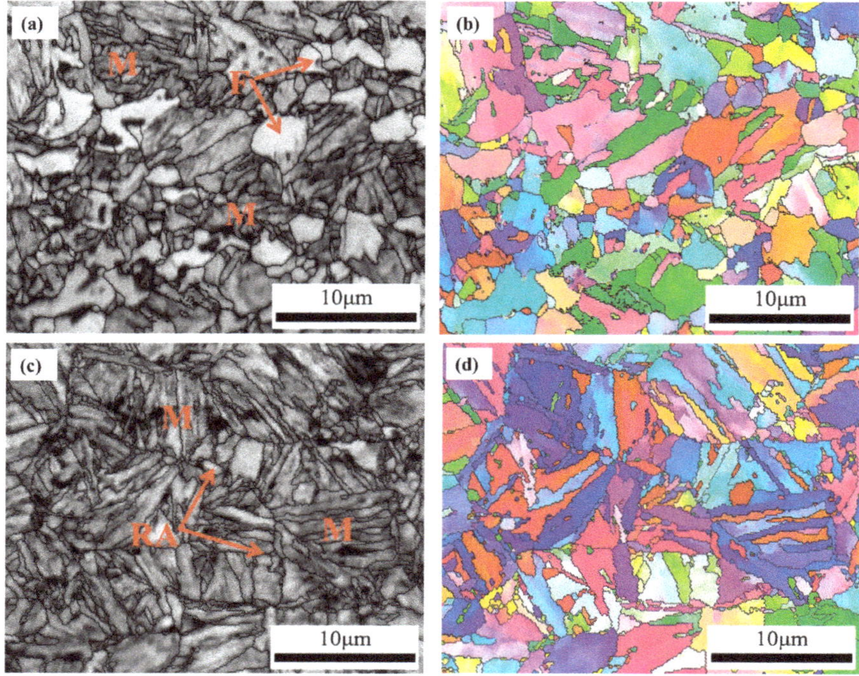

Figure 2. Microstructure of DP and QP steel prior to deformation. (**a**,**b**) Distribution of ferrite (F) and martensite (M) in DP steel. (**c**,**d**) Distribution of martensite (M) and retained austenite (RA) in QP steel.

3.2. Bending Springback Properties

The bending springback properties of the DP and QP steel are shown in Figure 3, as a function of bending angle. Figure 3a illustrated that the stress of the DP steel and QP steel increased rapidly with the bending angle. The DP steel yielded at a bending angle of around 10°, followed by significant bending deformation. However, the QP steel yielded at a bending angle of around 17°. Such differences in yielding are often found in unixial tension tests of automotive steel because of the different strengths [20,21]. Interrupted bending specimens, strained at 74°, 104°, and 128°, were conducted for the purpose of investigating the springback properties. Figure 3b shows the plot of springback angle versus bending angle in both DP and QP steel. The springback angle of the DP steel increased with the increase in the bending angle. This phenomenon is in accordance with the results of Peng et al. [22] and Li et al. [16]. Nevertheless, the springback angle of the QP steel strained at 104° was lower than that strained at 74°. After achieving a low value of 14.75° at the bending angle of 104°, the springback angle obtained a relatively high value of 18.05° when QP steel was strained at 128°. Moreover, the springback angles of the QP steel deformed at different angles were higher than that of DP steel. According to Gardiner [23] and Wollter [24], springback angle is dependent on yield strength and elastic modulus during bending. DP steel and QP steel have the same elastic modulus (202 GPa). Thus, the high springback angle of the QP steel was probably attributable to the high yield strength.

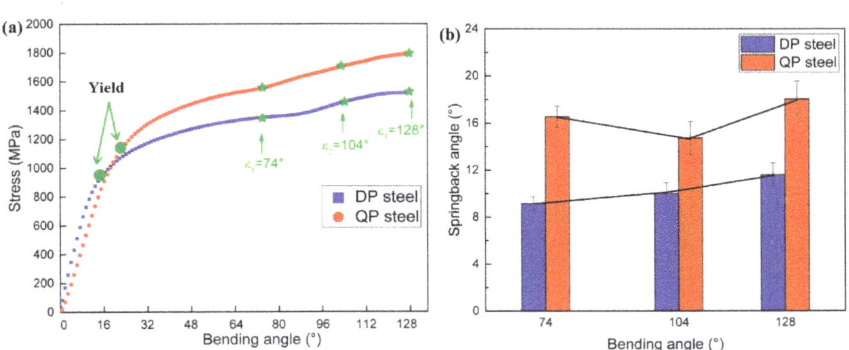

Figure 3. Bending springback properties of the DP and QP steel. (**a**) Stress-bending angle curves; (**b**) Springback angle of DP and QP steel strained at different bending angles.

3.3. Microstructure Characterization of the DP and QP Steel

3.3.1. Deformation Microstructure of the DP Steel

Microstructure characterization of the bending specimen at a 74° angle is compiled in Figure 4. Grain boundary maps (Figure 4a,c) of the EBSD measurements revealed that the majority of the angles of the grain boundary had values below 15° (low-angle grain boundary; LAGB; black) in the outer zone and the inner zone of the bending specimen. However, the specimen revealed the majority of the high-angle grain boundaries (HAGB; >15°; green and yellow) were in the center zone (Figure 4b). It is a known fact that the low angle grain boundary represents the degree of deformation [25,26]. Thus, the large misorientation difference in the grain boundary at different zones indicated that the strain heterogeneity occurred during the bending test. The strain heterogeneity of bending specimens is also illustrated in local misorientation distribution maps displayed in Figure 4. From Figure 4d–f, the local misorientations of the outer zone and inner zone were around 2°, while the local misorientation of the center zone was around 1°. Betanda et al. [27] reported that local misorientation distribution maps are an effective indicator of local strain in crystal materials. Therefore, the difference in the local misorientation also confirmed the occurrence of strain heterogeneity. In order to reliably characterize deformation of the DP steel during the bending test, the local misorientation data were used as a measure. Quantification results for the local misorientation are compiled in Figure 5a. The relative frequency of low-angle local misorientation (LALM, <1°) in the center zone was relatively higher than that of the inner zone and the outer zone, while the relative frequency of the high-angle local misorientation (HALM, ≥1°) was lower. This result also manifested the strain heterogeneity of specimens in the thickness direction. Moreover, strain heterogeneity also occurred in the samples strained at 104° and 128°.

To characterize the degree of strain heterogeneity, the difference in kernel average misorientation (KAM) was used [28,29]. Figure 5b presents the difference in kernel average misorientation when specimens deformed at different bending levels. In Figure 5b, Doc represents the KAM difference between the outer zone and the center zone; and D_{IC} is the KAM difference between the inner zone and the center zone. From Figure 5b, it could be seen that Doc and D_{IC} increased as the increasing bending angle, indicating that the degree of strain heterogeneity was increasing during bending. According to Zajkani et al. [30], the springback behavior is related to strain heterogeneity of different zones in thickness direction. A higher degree of strain heterogeneity corresponded to a higher tendency for springback. Thus, the springback angle of the DP steel increased with the increase in the bending angle, which corresponded well with the results in Figure 3b.

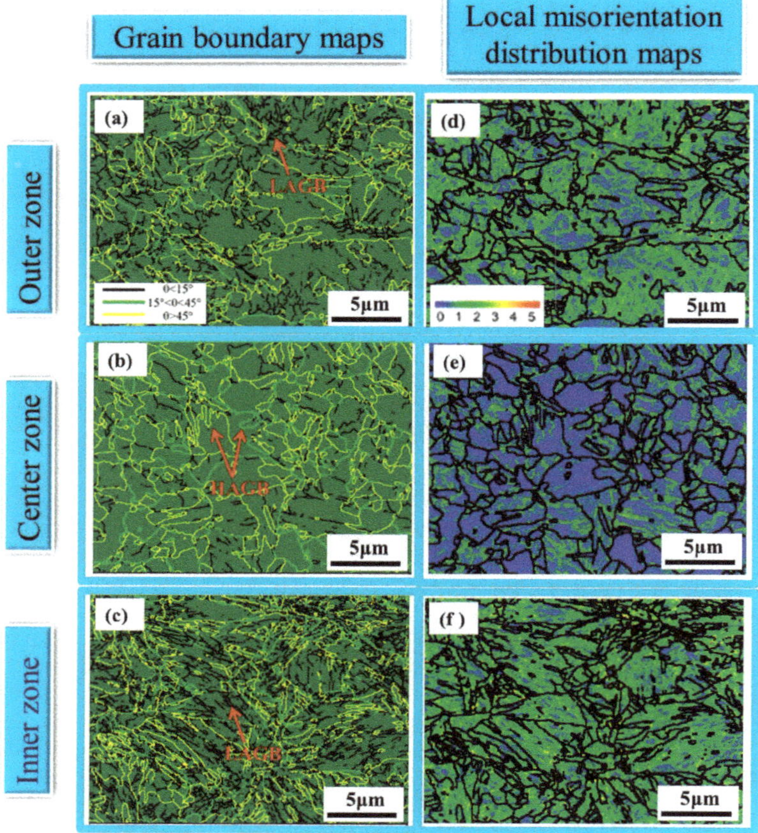

Figure 4. EBSD analysis for outer zone (**a,d**), center zone (**b,e**), and inner zone (**c,f**) of the bending specimen at 74° angle: EBSD grain boundary maps (left), and EBSD local misorientation distribution maps (right).

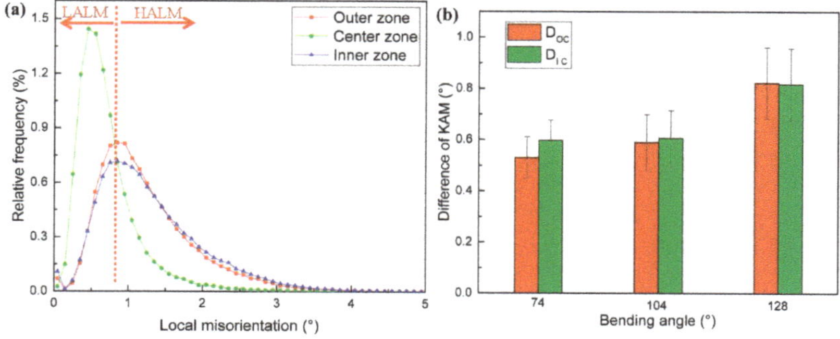

Figure 5. Change in the local misorientation and the difference in kernel average misorientation (KAM). (**a**) Change in the local misorientation of bending specimen at 74° angle; (**b**) Difference in kernel average misorientation in specimens strained at different bending angles.

3.3.2. Deformation Microstructure of the QP Steel

EBSD maps of QP specimens strained at 74° and 128° showed similar results to those of the DP steel. However, the QP sample deformed at a 104° angle exhibited different characterisations in microstructure (as illustrated in Figure 6a–f). The low-angle grain boundary contents of the outer zone and inner zone were similar to those of the center zone. Moreover, the local misorientations of the outer zone, center zone, and inner zones were all around 1°. The results indicated that the strain heterogeneity was relatively low. The local misorientation data in Figure 7a showed that the LALM and HALM frequency differences of the three zones were also low. Frequencies differences in LALM and HALM also confirmed the low strain heterogeneity at a 104° bending angle.

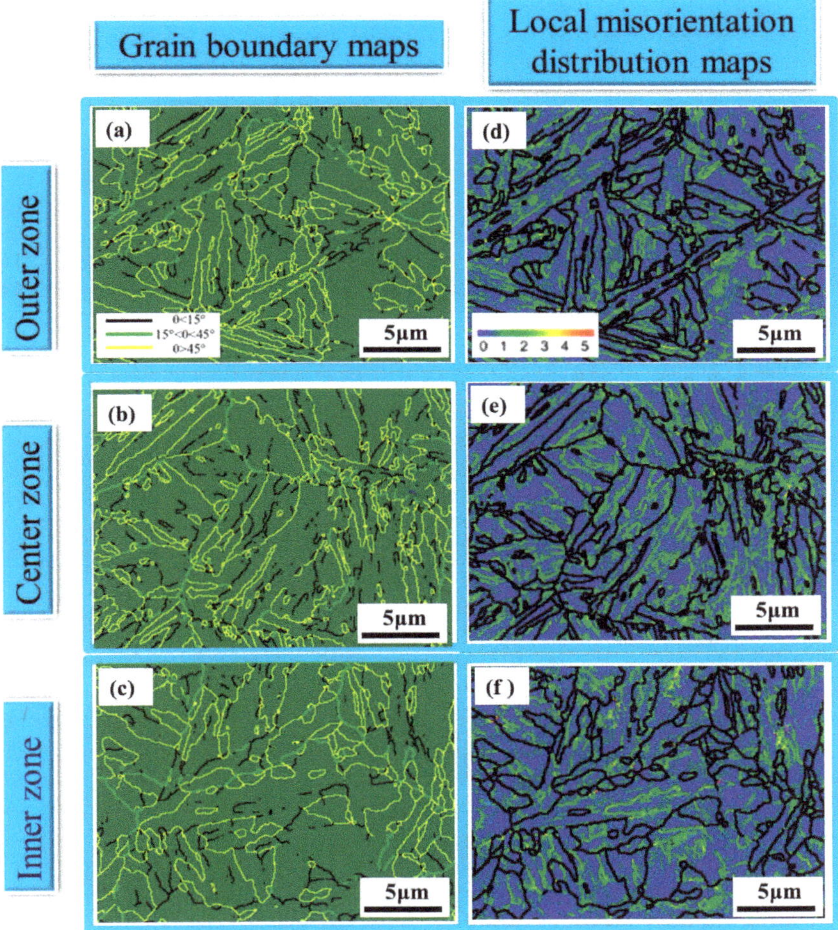

Figure 6. EBSD analysis for outer zone (**a**,**d**), center zone (**b**,**e**), and inner zone (**c**,**f**) of the bending specimen at 104° angle: EBSD grain boundary maps (left), and EBSD local misorientation distribution maps (right).

Analogous to the DP steel sample, the difference in kernel average misorientation was also used to characterize strain heterogeneity. Figure 7b displays the difference in kernel average misorientation of QP specimens strained at different bending levels. Unlike DP steel, the Doc and D_{I}c of QP steel obtained relatively low values at 104°. The lower values of Doc and D_{I}c represented the lower degree of strain heterogeneity. Moreover, the low degree of strain heterogeneity corresponded to the lower tendency for springback. Therefore, the springback angle of QP steel was low when specimens deformed at a 104° bending angle, which also corresponded well with the results in Figure 3b.

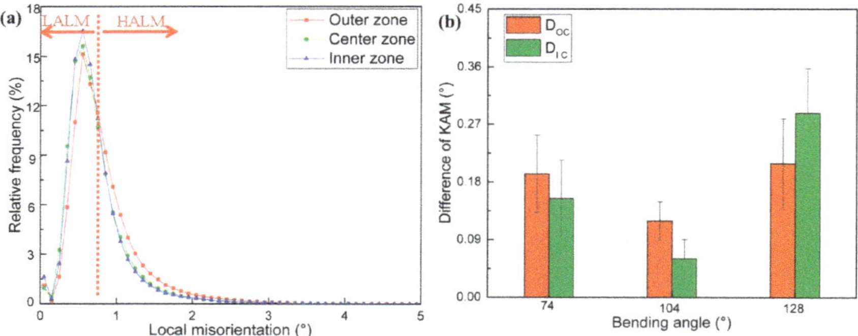

Figure 7. Change in local misorientation and difference in kernel average misorientation (KAM). (**a**) Change in local misorientation of bending specimen at 104° angle; (**b**) Difference in kernel average misorientation with specimens strained at different bending angles. LALM: low-angle local misorientation; HALM: high-angle local misorientation.

3.3.3. Transformation Characterization of Retained Austenite in the QP Steel

Figure 8 indicates the change in the retained austenite volume fractions at different bending angles. When the sample deformed at a 74° bending angle, the RA volume fractions in the outer zone, center zone, and inner zone were around 14%. Compared to the volume fraction of RA in specimens prior to bending (15.9%), only 1.9% retained austenite transformed when specimens strained at a 74° angle. It is a known fact that the RA transformation is affected by the surrounding phase [31,32]. In this research, the surrounding phase of retained austenite was only martensite due to the quenching and partitioning process. According to Song et al. [33], the retained austenite between martensite laths remained stable, while retained austenite within ferrite grains transformed preferentially during straining. Thus, the small transformation degree of retained austenite may be due to the protective effect of the surrounding martensite laths. As the sample was strained at a 104° bending angle, the retained austenite volume fraction of the outer zone and inner zone decreased significantly. The decrease in retained austenite indicated the deformation-induced martensitic transformation at higher strain. After the decrease in the outer zone and inner zone, the volume fraction of RA in the center zone decreased at a 128° bending angle. However, the RA volume fraction in the outer zone and the inner zone changed slightly. This indicated that a small degree of transformation occurred in the outer zone and the inner zone at this angle.

Figure 9 shows the retained austenite, twin martensite, and the orientation relationship between retained austenite and martensite. The film-like RA could be found between martensite laths (see Figure 9a). During deformation, the film-like RA can transform into twin martensite during deformation [33]. Nevertheless, twin martensite could not be seen at a 74° bending angle, as the lath martensite, having high strength, was not subjected to severe plastic deformation at this angle. With the sample was strained to 104°, twin martensite could be found within lath martensite (Figure 9b). Furthermore, the orientation relationship between retained austenite and martensite was a N-W relation: {1 1 1}γ//{1 1 0}M, <0 1 1>γ//<0 0 1>M, as seen in Figure 9c,d. The occurrence of twin martensite at a 104° bending angle corresponded to the decrease of RA volume fraction in Figure 8, indicating the deformation-induced martensitic transformation at higher strain.

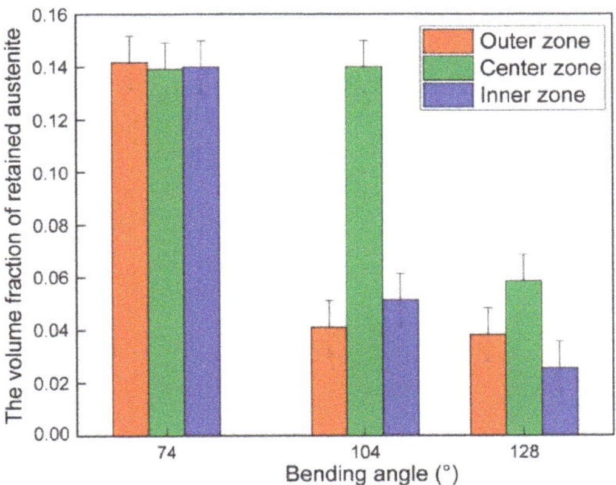

Figure 8. Change in the retained austenite volume fraction at different bending angles.

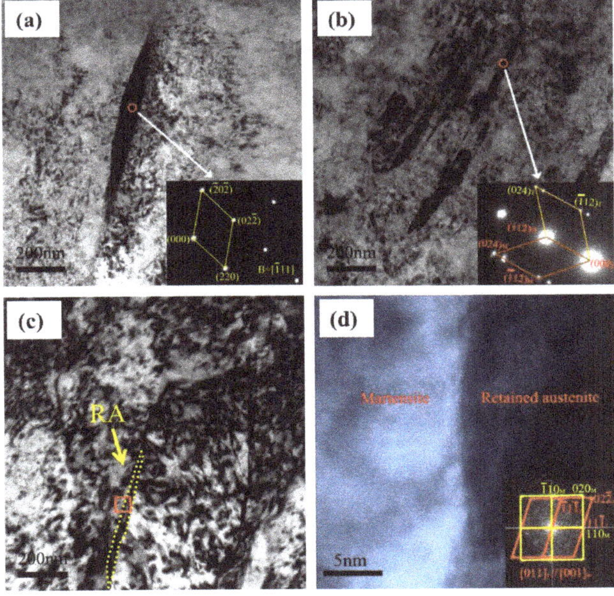

Figure 9. TEM images showing the retained austenite, twin martensite, and the orientation relationship between retained austenite and martensite. (**a**) Retained austenite and surrounding martensite; (**b**) twin martensite; and (**c**) the orientation relationship between retained austenite and martensite. (**d**) High-resolution TEM micrograph indicating N-W relation of the selected red rectangle region in (**c**).

3.4. Strain Heterogeneity and Transformation of Retained Austenite in QP Steel

As mentioned above, springback is related to the degree of strain heterogeneity that is highly dependent on the stress distribution [34]. According to Yang et al. [13] and Mishra et al. [34], the stress distribution of specimens in the thickness direction is outlined schematically in Figure 10a. During the bending process, the stress values in the thickness direction changed from "zero" at the center plane, to maximum at the inner and outer surface (indicated by red arrows in Figure 10a). However, during the bending process of the specimens, the middle surface is moved to the inner zone due to the radial force [23,24,35]. After the movement of middle surface, the stress distribution along the thickness changed, as suggested by the blue arrows in Figure 10a. Thus, three zones, namely the outer zone, center zone, and inner zone, were divided based on the stress difference. Higher stress in the outer zone and inner zone resulted in the higher strain (Figure 10b). However, the strain in the center zone was low during the bending process. Therefore, strain heterogeneity occurred in the thickness direction during the bending process. According to Zajkani et al. [30], a higher degree of strain heterogeneity corresponds to higher tendency for springback. Thus, the change trends of springback angles in both DP steel and QP steel were the same as that of the strain heterogeneity degree.

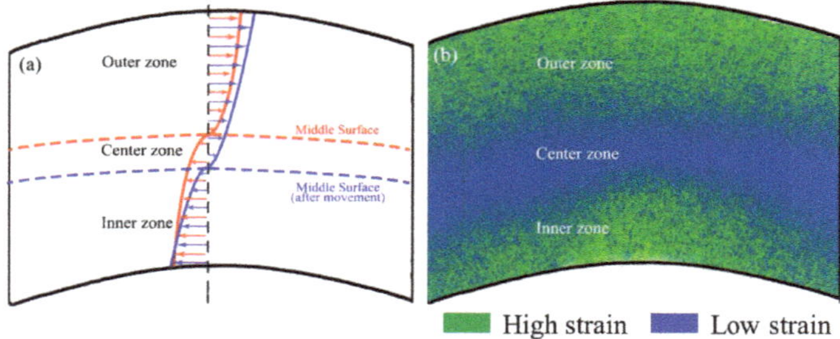

Figure 10. Schematic diagram of stress and strain distribution during bending process. (**a**) Stress distribution; (**b**) Strain distribution.

According to the change of the retained austenite volume fractions in Section 3.3.3, the retained austenite transformation during bending process is displayed schematically in Figure 11. As illustrated in Figure 11a, strain heterogeneity occurred during the bending process. However, the RA remained stable as the deformation did not reach a certain critical degree [36,37]. With further deformation, the retained austenite in the outer zone and inner zone transformed into martensite (as indicated in Figure 9). Moreover, the outer zone and inner zone obtain high strength due to the transformation of martensite [38,39], while the center zone has relatively low strength. Thus, the center zone began to deform due to the relatively low strength (Figure 11b). The deformation of the center zone led to low strain heterogeneity, resulting in the lower tendency for springback. When specimens were strained at a larger angle, the retained austenite of the center zone transformed into martensite (Figure 11c). After the transformation, the strength of the center zone was almost the same as the strength of the outer zone and inner zone. Thus, the deformation occurred in the outer zone and inner zone again. High strain heterogeneity took place and the springback angle was high at this bending level.

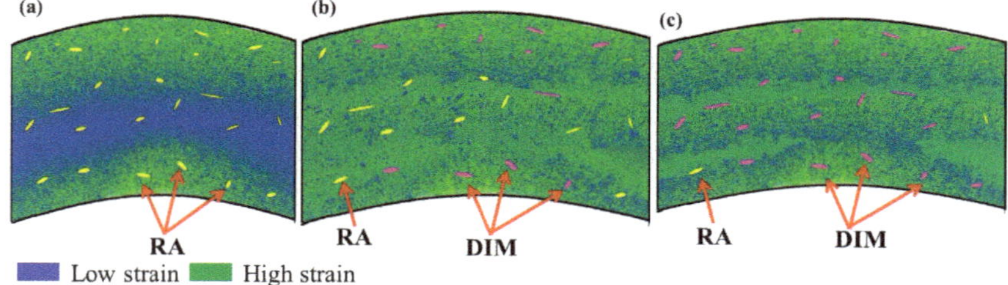

Figure 11. Schematic diagram of retained austenite transformation during the bending process. (**a**) QP specimens strained at low bending angle; (**b**) QP specimens strained at medium bending angle; (**c**) QP specimens strained at high bending angle. 'RA' and 'DIM' represent 'retained austenite' and 'deformation-induced martensite', respectively.

4. Conclusions

In this paper, the effect of the strain heterogeneity and transformation of metastable austenite on springback behavior was investigated. The conclusions can be summarized as follows:

(1) The increasing the bending angle from 74° increased the springback angle of DP steel. In contrast, the increasing the bending angle from 74° decreased the springback up to a certain angle and then it started to increase at higher bending angles. This direct comparison indicated that QP steel could possess lower springback at proper bending angles.

(2) During the bending process, strain heterogeneity occurred in the thickness direction of the DP steel and the QP steel. The strain heterogeneity degree of DP steel increased following an increasing bending angle, while the degree of strain heterogeneity of QP steel decreased at the bending angle of 104°. This phenomenon fitted well with the low springback angle when QP steel was strained at the 104° bending angle.

(3) There was preferential transformation of retained austenite in the outer zone and the inner zone of QP steel during the deformation process. Due to the deformation-induced martensitic transformation at a 104° bending angle, the deformation was more homogeneous, and the springback angle decreased.

Author Contributions: Y.Y. and Z.M. designed the experiments. S.L. and H.J. analyzed the mechanical properties data. Y.Y., H.L. and J.L. performed the EBSD measurements and analyzed the data. All authors contributed to the interpretation of the results and the writing of the final version of the manuscript.

Funding: This research was funded by National Key Research and Development Program of China (grant number: 2016YFB0101605 and 2017YFB0304404) and Shandong Provincial Natural Science Foundation of China (grant number: ZR2018MEM007).

Acknowledgments: Hai-Tao Jiang acknowledges financial support from the National Key Research and Development Program of China (2016YFB0101605). Zhen-Li Mi acknowledges financial support from the National Key Research and Development Program of China (2017YFB0304404). Hui Li acknowledges financial support from the Shandong Provincial Natural Science Foundation of China (Grant ZR2018MEM007).

Conflicts of Interest: The authors declare no conflict of interest.

References

1. Rehrl, J.; Mraczek, K.; Pichler, A.; Werner, E. Mechanical properties and fracture behavior of hydrogen charged AHSS/UHSS grades at high-and low strain rate tests. *Mater. Sci. Eng. A* **2014**, *590*, 360–367. [CrossRef]

2. Gyasi, E.A.; Kah, P.; Wu, H.; Kesse, M.A. Modeling of an artificial intelligence system to predict structural integrity in robotic GMAW of UHSS fillet welded joints. *Int. J. Adv. Manuf. Technol.* **2017**, *93*, 1139–1155. [CrossRef]

3. Mohrbacher, H. Property Optimization in As-Quenched Martensitic Steel by Molybdenum and Niobium Alloying. *Metals* **2018**, *8*, 234. [CrossRef]

4. Speer, J.; Matlock, D.K.; De Cooman, B.C.; Schroth, J.G. Carbon partitioning into austenite after martensite transformation. *Acta Mater.* **2003**, *51*, 2611–2622. [CrossRef]

5. Allain, S.Y.P.; Geandier, G.; Hell, J.C.; Soler, M.; Danoix, F.; Gouné, M. Effects of Q&P Processing Conditions on Austenite Carbon Enrichment Studied by In Situ High-Energy X-ray Diffraction Experiments. *Metals* **2017**, *7*, 232.

6. Santofimia, M.J.; Zhao, L.; Petrov, R.; Sietsma, J. Characterization of the microstructure obtained by the quenching and partitioning process in a low-carbon steel. *Mater. Characterization* **2008**, *59*, 1758–1764. [CrossRef]

7. Seo, K.Y.; Kim, J.H.; Lee, H.S.; Kim, J.H.; Kim, B.M. Effect of Constitutive Equations on Springback Prediction Accuracy in the TRIP1180 Cold Stamping. *Metals* **2017**, *8*, 18. [CrossRef]

8. Zang, S.; Sun, L.; Niu, C. Measurements of Bauschinger effect and transient behavior of a quenched and partitioned advanced high strength steel. *Mater. Sci. Eng. A* **2013**, *586*, 31–37. [CrossRef]

9.   Geng, L.; Wagoner, R.H. Role of plastic anisotropy and its evolution on springback. *Int. J. Mech. Sci.* **2002**, *44*, 123–148. [CrossRef]

10. Kim, H.; Kim, C.; Barlat, F.; Lee, M.G. Nonlinear elastic behaviors of low and high strength steels in unloading and reloading. *Mater. Sci. Eng. A* **2013**, *562*, 161–171. [CrossRef]

11. Kim, H.; Kimchi, M.; Kardes, N.; Altan, T. Effects of variable elastic modulus on springback predictions in stamping advanced high-strength steels (AHSS). *Steel Res. Int.* **2011**, *8*, 628–633.

12. Eggertsen, P.A.; Mattiasson, K. On constitutive modeling for springback analysis. *Int. J. Mech. Sci.* **2010**, *52*, 804–818. [CrossRef]

13. Yang, X.; Choi, C.; Sever, N.K.; Altan, T. Prediction of springback in air-bending of Advanced High Strength steel (DP780) considering Young's modulus variation and with a piecewise hardening function. *Int. J. Mech. Sci.* **2016**, *105*, 266–272. [CrossRef]

14. Wang, Z.; Hu, Q.; Yan, J.; Chen, J. Springback prediction and compensation for the third generation of UHSS stamping based on a new kinematic hardening model and inertia relief approach. *Int. J. Adv. Manuf. Technol.* **2017**, *90*, 875–885. [CrossRef]

15. Li, H.; Yang, H.; Song, F.F.; Zhan, M.; Li, G.J. Springback characterization and behaviors of high-strength Ti-3Al-2.5 V tube in cold rotary draw bending. *J. Mater. Process. Technol.* **2012**, *212*, 1973–1987. [CrossRef]

16. Li, H.; Yang, H.; Zhan, M.; Kou, Y.L. Deformation behaviors of thin-walled tube in rotary draw bending under push assistant loading conditions. *J. Mater. Process. Technol.* **2010**, *210*, 143–158. [CrossRef]

17. Zou, D.Q.; Li, S.H.; He, J.; Gu, B.; Li, Y.F. The Deformation Induced Martensitic Transformation and Mechanical Behavior of Quenching and Partitioning steels under Complex Loading Process. *Mater. Sci. Eng. A* **2018**, *715*, 243–256. [CrossRef]

18. Doege, E.; Kulp, S.; Sunderkötter, C. Properties and application of TRIP-steel in sheet metal forming. *Steel Res. Int.* **2002**, *73*, 303–308. [CrossRef]

19. VDA 238-100 Test Specification Draft: Plate Bending Test for Metallic Materials. 12/2010. Available online: https://www.vda.de/en/services/Publications/vda-238-100-plate-bending-test-for-metallic-materials.html (accessed on 20 June 2017).

20. Zhao, H.S.; Li, W.; Zhu, X.; Lu, X.H.; Wang, L.; Zhou, S.; Jin, X.J. Analysis of the relationship between retained austenite locations and the deformation behavior of quenching and partitioning treated steels. *Mater. Sci. Eng. A* **2016**, *649*, 18–26. [CrossRef]

21. Bayramin, B.; Şimşir, C.; Efe, M. Dynamic strain aging in DP steels at forming relevant strain rates and temperatures. *Mater. Sci. Eng. A* **2017**, *704*, 164–172. [CrossRef]

22. Xu, Z.; Peng, L.; Bao, E. Size effect affected springback in micro/meso scale bending process: Experiments and numerical modeling. *J. Mater. Process. Technol.* **2018**, *252*, 407–420. [CrossRef]

23. Gardiner, F.J. The springback of metals. *Trans. ASME* **1957**, *79*, 1–9.

24. Wollter, K. Freies Biegen von Blechen. *VDI-Forschungsh* **1952**, *435*, 11–15.

25. Gu, Y.; Xiang, Y.; Srolovitz, D.J. Relaxation of low-angle grain boundary structure by climb of the constituent dislocations. *Scr. Mater.* **2016**, *114*, 35–40. [CrossRef]

26. Read, W.T.; Shockley, W. Dislocation models of crystal grain boundaries. *Phys. Rev.* **1950**, *78*, 275–280. [CrossRef]

27. Betanda, Y.A.; Helbert, A.L.; Brisset, F.; Mathon, M.H.; Waeckerlé, T.; Baudin, T. Measurement of stored energy in Fe–48% Ni alloys strongly cold-rolled using three approaches: Neutron diffraction, Dillamore and KAM approaches. *Mater. Sci. Eng. A* **2014**, *614*, 193–198. [CrossRef]

28. Kwon, E.P.; Fujieda, S.; Shinoda, K.; Suzuki, S. Characterization of transformed and deformed microstructures in transformation induced plasticity steels using electron backscattering diffraction. *Mater. Sci. Eng. A* **2011**, *528*, 5007–5017. [CrossRef]

29. Kamaya, M. Assessment of local deformation using EBSD: Quantification of accuracy of measurement and definition of local gradient. *Ultramicroscopy* **2011**, *111*, 1189–1199. [CrossRef] [PubMed]

30. Zajkani, A.; Hajbarati, H. Investigation of the variable elastic unloading modulus coupled with nonlinear kinematic hardening in springback measuring of advanced high-strength steel in U-shaped process. *J. Manuf. Process.* **2017**, *25*, 391–401. [CrossRef]

31. Yang, Y.G.; Mi, Z.L.; Xu, M.; Xiu, Q.; Jiang, H.T. Impact of intercritical annealing temperature and strain state on mechanical stability of retained austenite in medium Mn steel. *Mater. Sci. Eng. A* **2018**, *725*, 389–397. [CrossRef]

32. Timokhina, I.B.; Hodgson, P.D.; Pereloma, E.V. Effect of microstructure on the stability of retained austenite in transformation-induced-plasticity steels. *Metall. Mater. Trans. A* **2004**, *35*, 2331–2341.

33. Song, C.; Yu, H.; Li, L.; Zhou, T.; Lu, J.; Liu, X. The stability of retained austenite at different locations during straining of I&Q&P steel. *Mater. Sci. Eng. A* **2016**, *670*, 326–334.

34. Mishra, A.; Thuillier, S. Investigation of the rupture in tension and bending of DP980 steel sheet. *Int. J. Mech. Sci.* **2014**, *84*, 171–181. [CrossRef]

35. Roumina, R.; Bruhis, M.; Masse, J.P.; Zurob, H.S.; Jain, M.; Bouaziz, O.; Embury, J.D. Bending properties of functionally graded 300M steels. *Mater. Sci. Eng. A* **2016**, *653*, 63–70. [CrossRef]

36. Olson, G.B.; Cohen, M. A mechanism for the strain-induced nucleation of martensitic transformations. *J. Less Common Met.* **1972**, *28*, 107–118. [CrossRef]

37. Stringfellow, R.G.; Parks, D.M.; Olson, G.B. A constitutive model for transformation plasticity accompanying strain-induced martensitic transformations in metastable austenitic steels. *Acta Metall. Mater.* **1992**, *40*, 1703–1716. [CrossRef]

38. He, B.B.; Hu, B.; Yen, H.W.; Cheng, G.J.; Wang, Z.K.; Luo, H.W.; Huang, M.X. High dislocation density–induced large ductility in deformed and partitioned steels. *Science* **2017**, *357*, 1029–1032. [CrossRef] [PubMed]

39. Kim, J.H.; Seo, E.J.; Kwon, M.H.; Kang, S.; De Cooman, B.C. Effect of Quenching Temperature on Stretch Flangeability of a Medium Mn Steel Processed by Quenching and Partitioning. *Mater. Sci. Eng. A* **2018**. [CrossRef]

Article

Effects of Cr and Mo on Mechanical Properties of Hot-Forged Medium Carbon TRIP-Aided Bainitic Ferrite Steels

Koh-ichi Sugimoto [1,*], Sho-hei Sato [2], Junya Kobayashi [3] and Ashok Kumar Srivastava [4]

[1] Department of Mechanical Systems Engineering, Graduate School of Science and Technology, Shinshu University, Nagano 380-8553, Japan

[2] Department of Production Engineering, Sato Press Co., Ltd., Toyota 473-0933, Japan; s-sato@satopress.com

[3] Department of Mechanical Engineering, Graduate School of Science and Engineering, Ibaraki University, Hitachi 316-8511, Japan; junya.kobayashi.jkoba@vc.ibaraki.ac.jp

[4] Department of Metallurgical Engineering, School of Engineering, OP Jindal University, Raigarh 496109, India; ashok.srivastava@opju.ac.in

* Correspondence: sugimot@shinshu-u.ac.jp; Tel.: +81-90-9667-4482

Received: 12 September 2019; Accepted: 27 September 2019; Published: 30 September 2019

Abstract: In this study, the effects of Cr and Mo additions on mechanical properties of hot-forged medium carbon TRIP-aided bainitic ferrite (TBF) steel were investigated. If 0.5%Cr was added to the base steel with a chemical composition of 0.4%C, 1.5%Si, 1.5%Mn, 0.5%Al, and 0.05%Nb in mass%, the developed steel achieved the best combination of strength and total elongation. The best combination of strength and impact toughness was attained by multiple additions of 0.5%Cr and 0.2%Mo to the base steel. The excellent combination of strength and impact toughness substantially exceeded those of quenched and tempered JIS-SCM420 and 440 steels, although it was as high as those of 0.2%C TBF steels with 1.0%Cr and 0.2%Mo. The good impact toughness was mainly caused by uniform fine bainitic ferrite matrix structure and a large amount of metastable retained austenite.

Keywords: hot-forging; microalloying; TRIP-aided bainitic ferrite steel; retained austenite; tensile property; impact toughness

---

1. Introduction

In the past decades, first-, second-, and third-generation cold- and hot-rolled advanced high-strength steels (AHSSs) have been developed in the world [1,2]. Ferrite-martensite dual-phase steels, TRIP-aided steels with polygonal ferrite matrix structure, and complex steel are categorized as the first-generation AHSSs [1]. High-Mn twinning-induced plasticity (TWIP) steels are known as the typical second-generation AHSSs [2]. The typical third-generation AHSSs are transformation-induced plasticity (TRIP)-aided bainitic ferrite (TBF) [3], bainitic ferrite/martensite (TBM) and martensite (TM) steels [4–7], quenching and partitioning (Q&P) steels [8–10], carbide-free bainitic (CFB) steels (or nano-structured bainitic steels) [11–13], and medium Mn (M-Mn) steels [14–16]. Cold rolled AHSSs of 980–1180 MPa grade with excellent cold formability have already been applied to automotive body in white and seat frame in order to reduce the weight and enhance the crush safety [17–19]. In addition, 1180 MPa hot-rolled AHSS has been successfully applied in truck cylinders for concrete mixer [20].

Low- and medium-carbon TBF, TBM, and TM steels [4–7] are produced by a similar heat treatment process to low- and medium-carbon Q&P [8–10], CFB steels [11–13], and martensite type M-Mn steels [15], except Q&P steel subjected to two step Q&P process and dual-phase type M-Mn steels [14,16]. The heat treatment consists of austenitizing and subsequent austemper or martempering.

Recently, an interesting project for weight reduction and size-down of automotive forging parts such as Powertrain components etc., "The Lightweight Forging Initiative", was implemented in Germany [21,22]. In this project, V-bearing precipitation-hardening ferritic-pearlitic steels and bainitic steels without heat treatment after hot-forging were used on behalf of quenched and tempered martensitic steels for the weight reduction and size-down. For further weight reduction, TBF, TBM, and TM steels [23–25] are also very attractive as well as Q&P [26,27], CFB [26,28–33], and martensite type M-Mn steels [15], because their steels possess excellent mechanical properties such as tensile strength, impact toughness, fatigue strength, and delayed fracture strength.

In order to develop a new hot-forged TBF steel, Sugimoto et al. [34,35] investigated the effects of hot forging in γ region and the subsequent austemper (FA) process on the microstructure and mechanical properties of TBF steels with chemical compositions of 0.4%C-1.5%Si-1.5%Mn and 0.4%C-1.5%Si-1.5%Mn-1.0%Al-0.05%Nb. They obtained the following interesting results:

(1)   The FA process refined the microstructure and increased the volume fraction of retained austenite with a decrease in its carbon concentration.
(2)   Good combination of yield strength and impact toughness was achieved when austemper was conducted at temperatures above M_s.

If the TBF steels are applied to relatively large forging parts, high hardenability may be required to obtain the mixed microstructure of bainitic ferrite and metastable retained austenite. In general, hardenability of the steel is improved by the addition of alloying elements such as Cr, Mo, Ni, Mn, B, etc. However, there is no research investigating the effects of hardenability on microstructure and mechanical properties in the hot-forged medium-carbon TBF steels.

In the present study, the effects of Cr and Mo additions on the microstructure and mechanical properties (such as tensile properties and impact toughness) of 0.4%C-1.5%Si-1.5%Mn-0.5%Al- 0.05%Nb TBF steels subjected to the FA process were experimentally investigated. The mechanical properties were related to the microstructural and retained austenite characteristics. In order to investigate the effects of carbon content, the mechanical properties were compared with those of hot-forged low-carbon TBF steels (0.2%C-1.5%Si-1.5%Mn-0.04%Al-0.05%Nb-(0–1.0)%Cr-(0–0.2)%Mo). In addition, the mechanical properties were compared with those of commercial JIS-SCM420 and 440 steels [36].

2. Experimental Procedure

Three 100 kg ingots of 100 mm in diameter were vacuum-melted and then hot-forged to 32 mm in diameter. Chemical compositions of the steel bars are shown in Table 1. Steel A is a base steel containing 0.40%C, 1.49%Si, 1.49%Mn, 0.49%Al, and 0.048%Nb. Al and Nb were added to the base steel to stabilize the retained austenite and refine the prior austenite grain size, respectively. Steel B was obtained by adding about 0.51% Cr to the Steel A. Steel C was produced by further adding of up to 0.20%Mo to the Steel B. For the Steels B and C, Cr and Mo were added to improve the hardenability. The martensite-start and -finish temperatures (M_S and M_f) of the steels were measured using a dilatometer (Thermecmaster-Z, Fuji Electronic Ind. Co., Osaka, Japan). The continuous cooling transformation (CCT) diagrams of Steels A–C are shown in Figure 1. To investigate the effects of carbon content, four kinds of low-carbon 1.5%Si-1.5%Mn steels with different Cr and Mo contents (Steels D–G) were used. In addition, commercial JIS-SCM420 (0.21%C-0.26%Si-0.86%Mn-1.10%Cr-0.16%Mo) and JIS-SCM440 (0.39%C-0.19%Si-0.68%Mn- 0.95%Cr-0.17%Mo) steels [36] were also used to clear the effects of Si content.

Table 1. Chemical composition [mass%], martensite-start temperature M_S [°C] and carbon equivalent C_{eq}, [mass%] of medium- and low-carbon steels used.

Steel	C	Si	Mn	P	S	Cr	Mo	Al	Nb	N	O	M_S	C_{eq}
A	0.40	1.49	1.49	<0.005	0.0021	-	-	0.49	0.048	0.0009	0.0006	325	0.710
B	0.43	1.50	1.52	<0.005	0.0023	0.51	-	0.49	0.052	0.0009	0.0005	318	0.846
C	0.42	1.47	1.51	<0.005	0.0019	0.50	0.20	0.48	0.052	0.0010	0.0007	310	0.883
D	0.20	1.52	1.50	0.004	0.0021	-	-	0.039	-	0.0011	0.0010	436	0.513
E	0.21	1.49	1.50	0.004	0.0019	0.50	-	0.040	0.05	0.0012	0.0012	426	0.622
F	0.21	1.49	1.50	0.004	0.0019	1.00	-	0.040	0.05	0.0013	0.0012	416	0.722
G	0.18	1.48	1.49	0.004	0.0029	1.02	0.20	0.043	0.05	0.0010	0.0015	404	0.744

In this study, the effects of Cr and Mo were replaced by the following carbon equivalent (C_{eq}).

$$C_{eq} = C + Si/24 + Mn/6 + Ni/40 + Cr/5 + Mo/4 + V/24, \tag{1}$$

where C, Si, Mn, Ni, Cr, Mo, and V represent content in mass% of individual alloying elements.

Square bars of 20 mm thickness, 32 mm width, and 80 mm length, milled from steel bars of φ32 mm, were held at 900 °C for 1200 s and then hot-forged in one stage up to a reduction strain of 50% using a 400-ton hot-forging machine, followed by austemper at 350 °C for 1000 s (Figure 2b). The austemper temperature corresponds to an optimum temperature for impact toughness of Steels A to C [35]. Die temperature and strain rate of hot-forging were about 350 °C and about 50%/s, respectively. Other bars were subjected to the only austemper without hot-forging (Figure 2a). Hereafter, this process is called conventional austemper (CA) process. The same heat treatment as Figure 2 was conducted for Steels D–G. In this case, the austemper temperature (350 °C) is lower than M_s (404–436 °C), differing from Steels A–C, because the temperature gives the best mechanical properties in Steels D–G [36]. Heat treatment of quenching in oil and then tempering at 200–600 °C for 3600 s was conducted to JIS-SCM420 and 440 steels.

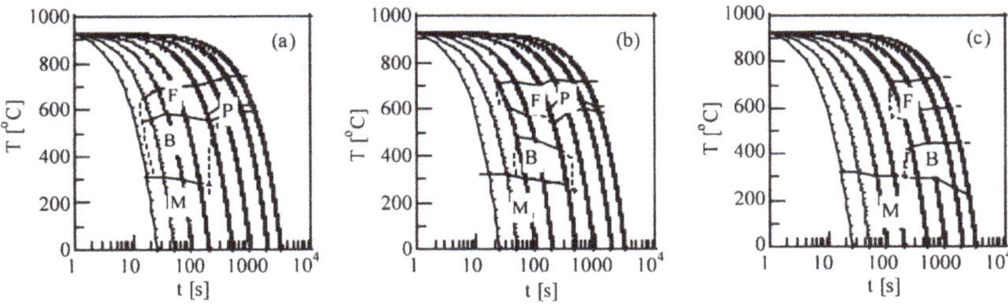

Figure 1. CCT diagrams of Steels (**a**) A, (**b**) B, and (**c**) C, in which "F", "P", "B", and "M" represent ferrite, pearlite, bainite, and martensite, respectively.

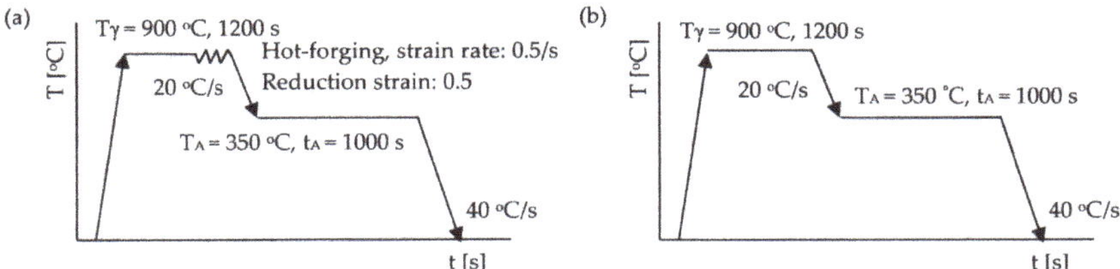

Figure 2. Schematic diagram of the (**a**) FA and (**b**) CA processes. $T\gamma$—austenitizing temperature; T_A—austempering temperature; t_A—holding time of austemper.

The microstructure of the steels was observed by field-emission scanning electron microscopy (FE-SEM; JSM-7000F, JEOL Ltd., Tokyo, Japan) using an instrument equipped with electron-backscatter diffraction (EBSD; OIM system, TexSEM Laboratories, Inc., Draper, UT, USA) system. The specimens for FE–SEM–EBSD analysis were milled with alumina powder and colloidal silica and then ion-thinned. In this case, some samples were mounted by resin and polished together to obtain the same surface condition.

The retained austenite characteristics of the steels were evaluated by X-ray diffractometry (XRD; RINT2000, Rigaku Co., Tokyo, Japan). The surfaces of the specimens were electropolished after being ground with Emery paper (#1200). The volume fraction of retained austenite phase ($f\gamma$, vol.%) was quantified from the integrated intensity of (200)α, (211)α, (200)γ, (220)γ, and (311)γ peaks obtained using Mo-Kα radiation [37]. The carbon concentration (C_γ, mass%) was estimated from the following empirical equation proposed by Dyson and Holmes [38]. Lattice constant (a_γ, 10^{-1} nm) was determined from the (200)γ, (220)γ, and (311)γ peaks of Cu-Kα radiation.

Mechanical stability of the retained austenite was calculated using the strain-induced transformation factor (k) defined by the following Equation [4]:

$$k = (\ln f\gamma_0 - \ln f\gamma)/\varepsilon, \tag{2}$$

where $f\gamma_0$ is an original volume fraction of retained austenite and $f\gamma$ is the volume fraction of retained austenite after being strained to plastic strain ε.

Tensile specimens of JIS-14B-type (22 mm length, 6 mm width, and 1.2 mm thickness) and JIS-4-type of sub-sized Charpy impact specimens with V-notch of 2 mm depth and 2.5 mm thickness were machined from 1/4 part of hot-forged plate thickness (Figure 3). Tensile tests were carried out at 25 °C using a tensile testing machine (AG-10TD, Shimadzu Co., Kyoto, Japan) under a crosshead speed of 1 mm/min. Impact tests were conducted at 25 °C using a conventional Charpy impact testing machine (CI-300, Tokyo Testing Machines Inc., Tokyo, Japan). Both tests were carried out using two specimens each.

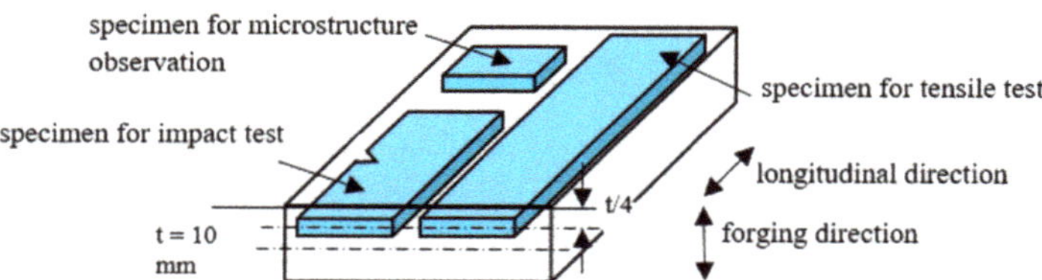

Figure 3. Sampling scheme of specimens from hot-forged plates. Unit of dimension is mm.

3. Results and Discussion

3.1. Microstructure and Retained Austenite Characteristics

Figure 4 shows EBSD phase maps of Steels A–C subjected to the CA and FA processes. It is found that the FA process considerably refines the sizes of prior austenitic grain, proeutectoid ferrite, bainitic ferrite, and retained austenite in Steels A–C. The higher the carbon equivalent of the steels, the more the volume fraction of acicular bainitic ferrite in the hot-forged steels. The microstructure of Steel C subjected to the FA process consists of acicular bainitic ferrite and retained austenite, with a negligible amount of proeutectoid ferrite and blacklike phase (Figure 4c,f). Refined, retained austenite phases distribute uniformly, compared with those of steel C without hot-forging. In this case, filmy- and particulate-retained austenite phases exist along the bainitic ferrite lath boundary and prior austenitic grain boundary, respectively. As the blacklike phases consist of many ultra-fine lath structures, as shown in Figure 4g–i, they are estimated to be a mixture of martensite and austenite or MA phase.

In hot-forged Steels C, the MA phase is slightly refined. The volume fraction of MA phase seems to be decreased by hot-forging.

The microstructure of Steel B subjected to the CA and FA processes seems to be complex. Considering the CCT diagram (Figure 1b), the matrix microstructure of Steel B subjected to the FA process consists of proeutectoid ferrite and acicular bainitic ferrite because of low carbon equivalent, although lath size of the acicular bainitic ferrite is larger than that of hot-forged Steel C. In the matrix structure, retained austenite and a small amount of refined MA phase are included. In Steel B without hot-forging, granular and acicular bainitic ferrites coexist in the matrix. As Steel A is characterized by the lowest hardenability, as shown in CCT diagram of Figure 1a, the matrix structures of Steel A subjected to the CA and FA processes are composed of much proeutectoid ferrite and bainitic ferrite (Figure 4a,d). Most of the bainitic ferrite seems to be granular. The retained austenite size is the largest of Steels A to C. The FA process increases MA phase fraction.

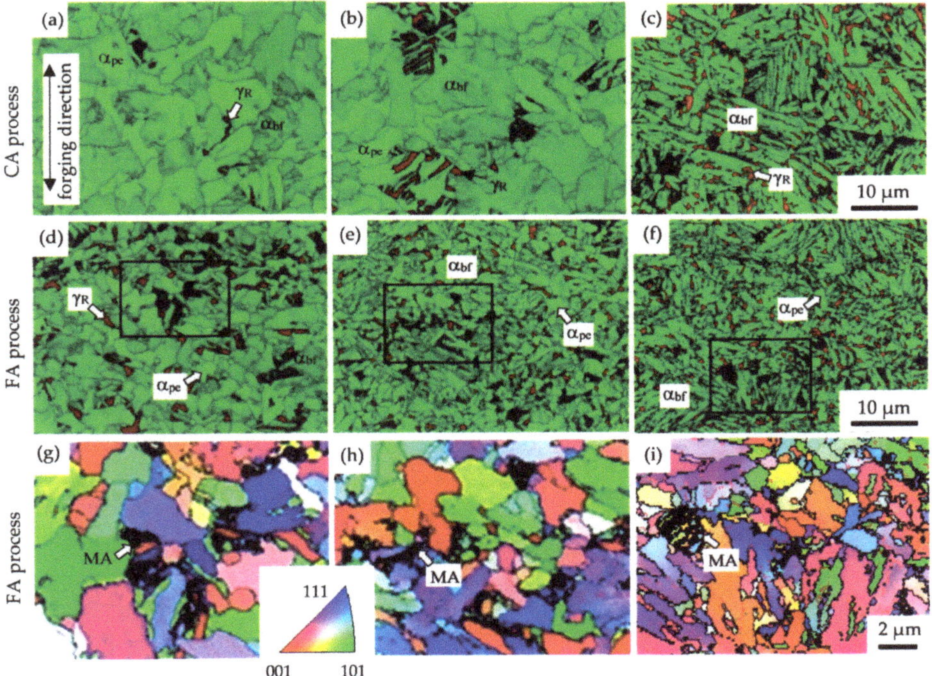

Figure 4. EBSP phase maps of Steels A (**a,d,g**), B (**b,e,h**) and C (**c,f,i**) subjected to CA and FA process. (**g–i**) are high magnification orientation maps of squares in (**d–f**), respectively. "α_{pe}", "α_{bf}", "γ_R" and "MA" represent pro-eutectoid ferrite, acicular bainitic ferrite, retained austenite (red) and martensite-austenite phase, respectively.

In general, image quality (IQ) index of the phase map by EBSD analysis can be related to dislocation density of individual phases [39]. Figure 5 shows IQ indices of FCC and BCC phases of Steels A and C without and with hot-forging. Peak IQ indices of both phases in Steel A subjected to the FA process are nearly equal to those after the CA process. This indicates that recrystallization after hot-forging finishes perfectly in both phases of Steel A. On the other hand, the peak IQ indices of both phases in Steel C shift to low index by FA process, different from Steel A. This indicates that bainitic ferrite and retained austenite phases in hot-forged Steel C inherit, to some extent, high dislocation density resulting from hot-forging via interaction between ultra-fine carbonitrides of Cr, Mo, and/or Nb. This should be further investigated using TEM analysis.

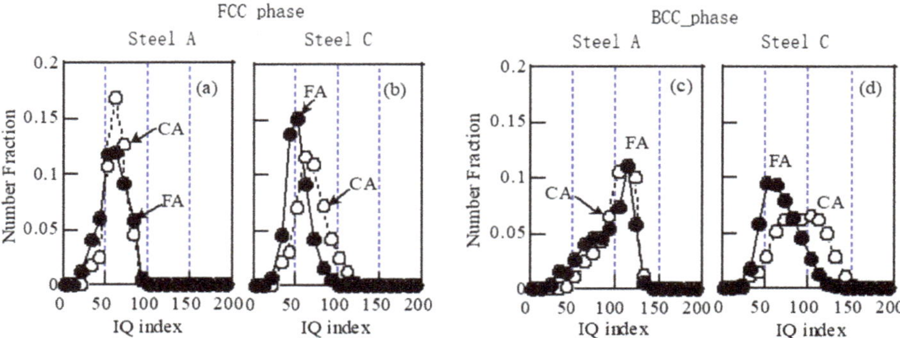

Figure 5. Image quality (IQ) distribution of (**a,b**) FCC and (**c,d**) BCC phase in Steels A and C subjected to CA and FA processes.

Figure 6a–d show an initial volume fraction ($f\gamma_0$), carbon concentration ($C\gamma_0$), and total carbon concentration ($f\gamma_0 \times C\gamma_0$) of retained austenite and a ratio of $f\gamma_0 \times C\gamma_0$ to added C content ($f\gamma_0 \times C\gamma_0/C$), as a function of carbon equivalent in Steels A to C. The volume fraction and carbon concentration increase with increasing carbon equivalent in hot-forged Steels A to C, as well as the total carbon concentration. The FA process tends to decrease the carbon concentration and total carbon concentration of retained austenite in comparison to CA process, although the FA process slightly increases the volume fraction of retained austenite except for Steel B. The $f\gamma_0 \times C\gamma_0/C$, which means a percentage solute carbon dissolved in retained austenite, increases with increasing carbon equivalents in Steels A to C subjected to FA process, although the FA process decreases the values.

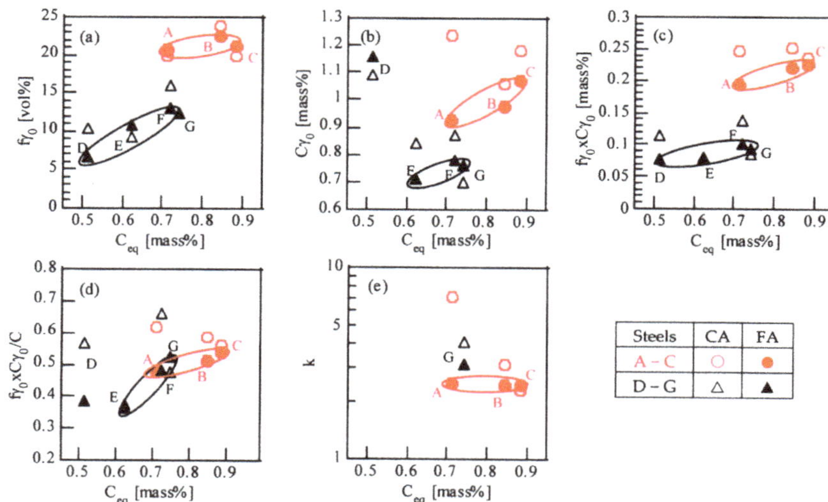

Figure 6. Carbon equivalent (C_{eq}) dependences of (**a**) initial volume fraction ($f\gamma_0$), (**b**) initial carbon concentration ($C\gamma_0$), (**c**) initial total carbon concentration ($f\gamma_0 \times C\gamma_0$) of retained austenite, (**d**) a ratio of $f\gamma_0 \times C\gamma_0$ to added C content ($f\gamma_0 \times C\gamma_0/C$), and (**e**) apparent k-value in Steels A–C and D–G subjected to the CA and FA processes.

Figure 6e shows apparent *k*-values of Steels A to C. The FA process tends to make the *k*-values smaller, particularly in Steel A, despite the low carbon concentration of retained austenite. This result indicates that the mechanical stability of retained austenite is principally decided by retained austenite size, compared with the carbon concentration, because the refined retained austenite suppresses the strain-induced martensite transformation [35,36]. It is found that Steels A to C subjected to the FA process apparently exhibit same *k*-values (or same mechanical stability of retained austenite). As the *k*-value of Steel C is evaluated at higher flow stress, the true mechanical stability is the highest of Steels A to C.

When the retained austenite characteristics of Steels A to C were compared with those of Steels D to G, Steels D to G were found to have lower carbon content, volume fraction, carbon concentration, and mechanical stability of the retained austenite than Steels A to C. It is noteworthy that carbon equivalent dependences of the retained austenite characteristics in Steels D to G are nearly the same as those of Steels A to C.

3.2. Tensile Properties

Figure 7 shows flow curves of Steels A, B, and C subjected to the CA and FA processes. Figure 8a–d show the tensile properties as a function of carbon equivalent. All steels exhibit continuous yielding and subsequent large strain hardening, especially in Steels A and C subjected to FA process. Yield stress or 0.2% offset proof stress (YS), tensile strength (TS), and yield ratio (YR = YS/TS) increase with increasing carbon equivalent in the steels subjected to the CA and FA processes. The FA process increases these strengths and decreases the yield ratio due to the increased strain-hardening rate.

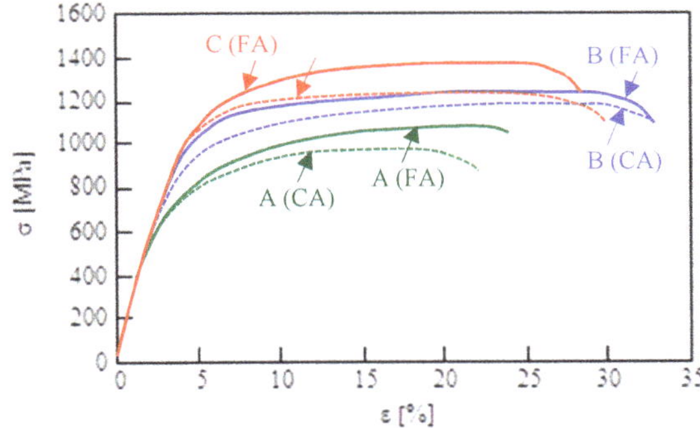

Figure 7. Nominal stress–strain curves of Steels A (green curves), B (blue curves) and C (red curves) subjected to CA and FA processes.

Carbon equivalent dependence of total elongation (TEl) is indistinct in Steels A to C subjected to the CA and FA processes. However, if Steel A is excluded, the total elongation decreases with increasing carbon equivalent, contrary to yield stress and tensile strength. Regarding total elongation, the FA process decreases the total elongation in Steel C depending on the trade-off relationship with tensile strength, although a difference in total elongations between the CA and FA processes is relatively small. As shown in Figure 8a,b, the largest combinations of yield stress and total elongation (YS × TEl), and tensile strength and total elongation (TS × TEl), are achieved in hot-forged Steel B. When these combinations were compared to those of hot-forged Steels D–G and JIS-SCM420 and 440 steels, hot-forged Steel B exhibits higher combinations than those of comparable steels, as well as hot-forged Steel C. In general, MA phase deteriorates the total elongation as an initiation site of void or crack. However, it is well known that the strain-induced transformation of retained austenite suppresses the void and/or crack initiation through the plastic relaxation effect in TBF steel [23]. Thus, it is considered that the MA phases do not decrease the total elongation.

According to Sugimoto et al. [35,36], tensile properties of hot-forged low- and medium-carbon TBF steels are influenced by the size and type of matrix structure, prior austenitic grain size, second phase properties, and retained austenite characteristics. In the present study, volume fraction of acicular bainitic ferrite increased with increasing carbon equivalent in Steels A to C (Figure 4). As the acicular bainitic ferrite is stronger than granular bainitic ferrite and proeutectoid ferrite, positive equivalent carbon dependences of yield stress and tensile strength (Figure 8a,b) may be mainly associated with the increased volume fraction of acicular bainitic ferrite. The increases in yield stress and tensile strength by hot-forging is supposed to be mainly caused by refined microstructure. High dislocation density of acicular bainitic ferrite and retained austenite (Figure 5) contributes to the highest yield stress and tensile strength obtained in Steel C.

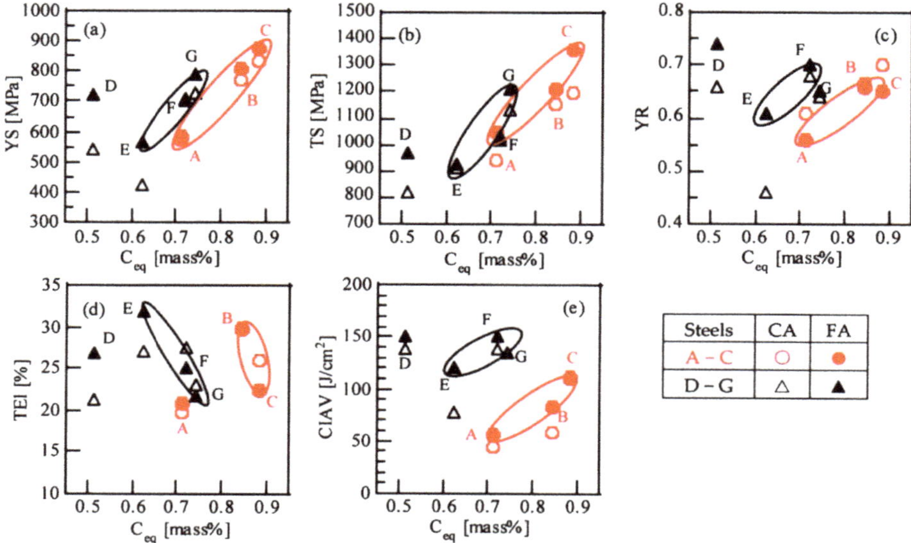

Figure 8. Carbon equivalent (C_{eq}) dependences of (**a**) yield stress (YS), (**b**) tensile strength (TS), (**c**) yield ratio (YR = YS/TS), (**d**) total elongation (TEl), and (**e**) Charpy impact absorbed value (CIAV) in Steels A−C and D−G subjected to the CA and FA processes.

The maximum total elongation and combinations YS × TEl and TS × TEl were achieved in Steel B subjected to the FA process (Figures 8d and 9a,b). Steel B possessed a dual-phase structure of proeutectoid ferrite and acicular bainitic ferrite. The dual-phase structure develops a compressive long-range internal stress in soft proeutectoid ferrite, differing from a single phase of acicular bainitic ferrite in Steel C. Resultantly, the internal stress may keep high strain-hardening rate in a large strain range and suppresses the diffuse necking, as well as an increase in flow stress [40]. Likewise, a large amount of metastable retained austenite enhances the strain-hardening rate via the TRIP effect and as a hard phase, especially in a large strain range. Therefore, excellent ductility of Steel B is considered to be brought by the dual-phase structure and a large amount of metastable retained austenite.

In Figure 9a,b, the combinations YS × TEl and TS × TEl of Steel B subjected to FA process was higher than those of Steels D−G subjected to FA process, as well as SCM420 and 440 steels. This may be associated with high retained austenite fraction and mechanical stability (Figure 6a,e). It is noteworthy that the combinations YS × TEl and TS × TEl of Steels D−G are of the same degree regardless of Cr and Mo contents.

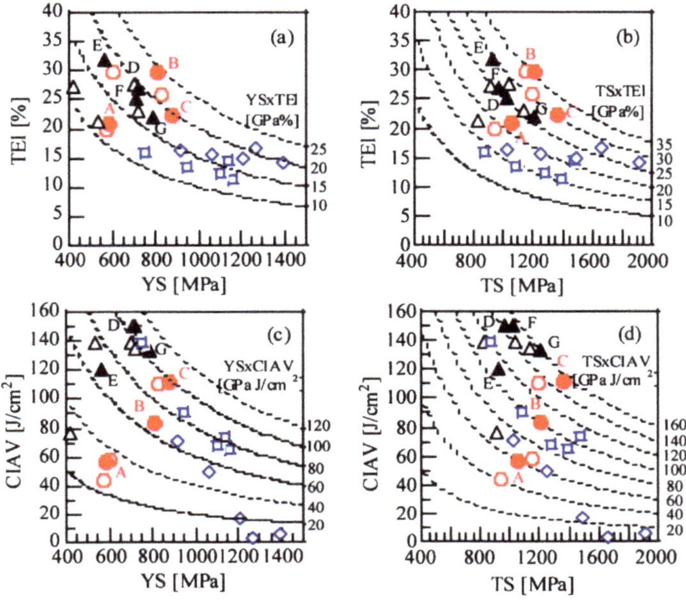

Figure 9. Combinations of (**a**,**b**) total elongation (TEl) with yield stress (YS) and with tensile strength (TS), respectively, and (**c**,**d**) V-notch Charpy impact absorbed value (CIAV) with YS and with TS, respectively, in Steels A–C (○●) and Steels D–G (△▲) subjected to the CA (open marks) and FA (solid marks) processes and quenched and tempered SCM420 (□) and 440 steels (◇).

3.3. Impact Toughness

Figure 8e shows Charpy impact absorbed value (CIAV) as a function of carbon equivalent in Steels A to C subjected to the CA and FA processes. The CIAV increases with increasing carbon equivalent in hot-forged Steels A to C. The FA process increases the CIAVs. The CIAVs are much lower than those of Steels D–G. Figure 9c,d show the combinations of YS and CIAV (YS × CIAV) and TS and CIAV (TS × CIAV) in Steels A–C subjected to the CA and FA processes. Maximum combinations are achieved in Steel C. The maximum combinations are the same extent as those of Steels F and G and are much higher than those of SCM420 and 440 steels. It is noteworthy that the single addition of Cr is very effective to enhance the combinations in low-carbon TBF steels (Steels D to G), differing from medium-carbon steels (Steels A to C).

Figure 10 shows SEM images of ductile fracture region and optical micrographs of fracture surface appearance in Steels A–C subjected to the CA and FA processes. High percent ductile fracture (PDF) can be related with high CIAV of these steels. Namely, the PDF increases with increasing CIAV, except Steel B subjected to the CA process. The FA process increases the PDF. It is found in the figure that the FA process makes the void size fine and void distribution uniform. As the PDF values of Steels A to C subjected to the FA process are 100%, ductile–brittle transition temperatures of all the steels are indicated to be lower than room temperature, as well as Steel C subjected to the CA process.

As with the ductility, impact toughness of low- and medium-carbon TBF steels is affected by size and type of matrix structure, prior austenitic grain size, second-phase properties, and retained austenite characteristics, which control the initiation and growth behavior of void and/or crack [23–25,41]. In this case, metastable retained austenite plastically relaxes the localized stress concentration through expansion strain on the strain-induced martensite transformation. Simultaneously, the strain-induced martensite makes a defense against crack initiation and propagation and brings in high-impact energy at an initial stage (strain-hardening stage) [41]. On the other hand, carbide-free acicular bainitic ferrite lath structure, a large amount of metastable retained austenite, and refined prior austenitic grain boundary suppress the initiation and growth of void or crack by lowering the localized stress concentration. This resultantly brings in high-impact energy at a final stage from top stress to fracture. In this case, carbon-enriched retained austenite in MA phases may also contribute to the impact energy through the plastic relaxation [23]. These microstructural features also lower the ductile–brittle transition temperature. From the above theory and the results of Figures 4 and 6e, high CIAV and combinations YS × CIAV and TS × CIAV of hot-forged Steel C may be mainly caused by refined uniform acicular bainitic ferrite matrix and a large amount of metastable retained austenite. Low CIAVs of Steels A and B may be associated with the existence of much proeutectoid ferrite, because the volume fraction and mechanical stability of the retained austenite of Steels A to C are nearly the same. The lowest CIAV of hot-forged Steel A is due to relatively coarse microstructure, because it plays as a stress concentration site.

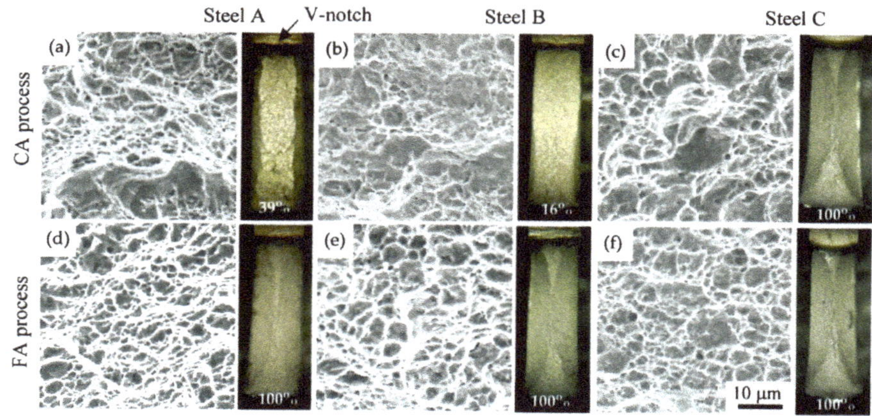

Figure 10. Typical SEM images of ductile fracture region at notch tip and optical micrographs of fracture surface appearance in Steels A (**a,d**), B (**b,e**), and C (**c,f**) subjected to the CA and FA processes. The numerals in the figure represent percent ductile fracture.

It is well known that impact toughness of medium-carbon steel considerably decreases compared to that of low-carbon steel [42]. In spite of medium-carbon content of 0.4%, Steel C possessed a relatively high CIAV (Figure 8e). Resultantly, Steel C achieved as high combinations YS × CIAV and TS × CIAV as Steels F and G with 0.2%C (Figure 9c,d). This may be caused by higher volume fraction and mechanical stability of retained austenite than those of Steels F and G (Figure 6), as well as carbide-free bainitic ferrite matrix structure. It is well known that Hy-tuf steel (0.24%C-1.42%Si-1.35%Mn-0.31%Cr-1.71%Ni-0.40%Mo-0.16%Cu) with 15–18 vol% retained austenite—which is usually applied to the aircraft parts, mining equipment, and other high-performance products—possesses high impact toughness [43]. However, the combinations YS × CIAV and TS × CIAV of the Hy-tuf steel are lower than those of hot-forged Steel C. This may be owing to the decreased carbon concentration (or low mechanical stability) of retained austenite.

4. Summary

The effects of Cr and Mo on the microstructure and mechanical properties of hot-forged 0.4%C-1.5%Si-1.5%Mn TBF steel were investigated. The main results are summarized as follows:

(1) Multiple additions of Cr and Mo developed a mixed structure of uniform bainitic ferrite matrix and a large amount of retained austenite in the TBF steel, with a negligible amount of proeutectoid

ferrite and MA phase. Hot-forging refined the mixed structure and increased the volume fraction and mechanical stability of retained austenite. The single addition of Cr reduced the proeutectoid ferrite fraction in the mixed structure, and it achieved the maximum retained austenite fraction.

(2) Multiple additions of Cr and Mo increased yield stress and tensile strength in the TBF steels. Hot-forging furthered the yield stress and tensile strength. The single addition of Cr achieved the maximum total elongation and combinations YS × TEl and TS × TEl because of a dual-phase structure of proeutectoid ferrite and acicular bainitic ferrite and high retained austenite fraction. In this case, multiple additions of Cr and Mo achieved slightly lower combinations than the single addition of Cr.

(3) Multiple addition of Cr and Mo brought on the maximum impact toughness and combinations YS × CIAV and TS × CIAV in hot-forged TBF steels. The combinations were the same as those of low-carbon TBF steels with chemical composition of 0.2%C-1.5%Si-1.5%Mn- 0.5%Cr-(0−0.2)%Mo. The excellent impact toughness was principally caused by uniform fine acicular bainitic ferrite matrix structure and a large amount of metastable retained austenite. It was considered that the mechanically stable retained austenite is especially effective to improve the impact toughness due to (i) the plastic relaxation of localized stress concentration and (ii) the increased martensite fraction via the strain-induced martensite transformation.

Author Contributions: Microstructure examination, tensile tests and impact tests were mainly performed by S.S. and J.K., the first draft of the paper was written by K.S. and A.K.S. All authors contributed to editing the paper, with final edits by K.S. and A.K.S.

Funding: This research was supported by a Grant-in-Aid for Scientific Research (B), The Ministry of Education, Science, Sports and Culture, Japan (No. 201325289262).

Acknowledgments: We thank Goro Arai from Nomura Unison Co., Ltd. for hot-forging tests.

Conflicts of Interest: The authors declare no conflict of interest.

References

1. Rana, R.; Singh, S.B. (Eds.) *Automotive Steels—Design, Metallurgy, Processing and Applications*; Woodhead Publishing: Cambridge, UK, 2016.
2. Zhao, J.; Jiang, Z. Thermomechanical processing of advanced high strength steels. *Prog. Mater. Sci.* **2018**, *94*, 174–242. [CrossRef]
3. Zackay, V.F.; Parker, E.R.; Fahr, D.; Bush, R. The enhancement of ductility in high-strength steels. *Trans. Am. Soc. Met.* **1967**, *60*, 252–259.
4. Sugimoto, K.; Murata, M.; Song, S. Formability of Al-Nb bearing ultrahigh-strength TRIP-aided sheet steels with bainitic ferrite and/or martensite matrix. *ISIJ Int.* **2010**, *50*, 162–168. [CrossRef]
5. Kobayashi, J.; Ina, D.; Yoshikawa, N.; Sugimoto, K. Effects of the addition of Cr, Mo and Ni on the microstructure and retained austenite characteristics of 0.2% C–Si–Mn–Nb ultrahigh-strength TRIP-aided bainitic ferrite steels. *ISIJ Int.* **2012**, *52*, 1894–1901. [CrossRef]
6. Kobayashi, J.; Song, S.; Sugimoto, K. Ultrahigh-strength TRIP-aided martensitic steels. *ISIJ Int.* **2012**, *52*, 1124–1129. [CrossRef]
7. Pham, D.V.; Kobayashi, J.; Sugimoto, K. Effects of microalloying on stretch-flangeability of TRIP-aided martensitic sheet steel. *ISIJ Int.* **2014**, *54*, 1943–1951. [CrossRef]
8. Speer, J.G.; Edmonds, D.V.; Rizzo, F.C.; Matlock, D.K. Partitioning of carbon from supersaturated plates of ferrite with application to steel processing and fundamentals of the bainitic transformation. *Curr. Opin. Solid State Mat. Sci.* **2004**, *8*, 219–237. [CrossRef]
9. Wang, L.; Speer, J.G. Quenching and partitioning steel heat treatment. *Metallogr. Microstruct. Anal.* **2013**, *2*, 268–281. [CrossRef]
10. Kähkönen, J.; Pierce, D.T.; Speer, J.G.; De Moor, E.; Thomas, G.A.; Coughlin, D.; Clarke, K.; Clarke, A. Quenched and partitioned CMnSi steels containing 0.3 wt.% and 0.4 wt.% carbon. *JOM* **2016**, *68*, 210–214.
11. Bhadeshia, H.K.D.H. Some phase transformations in steels. *Mater. Sci. Technol.* **1999**, *15*, 22–29. [CrossRef]

12. Garcia-Mateo, C.; Caballero, F.G.; Sourmail, T.; Smanio, V.; De Andres, C.G. Industrialised nanocrystalline bainitic steels—Design approach. *Int. J. Mater. Res.* **2014**, *105*, 725–734. [CrossRef]

13. Zhang, X.; Xu, G.; Wang, X.; Embury, D.; Bouaziz, O.; Purdy, G.P.; Zurob, H.S. Mechanical behavior of carbide-free medium carbon bainitic steels. *Metall. Mater. Trans. A* **2014**, *45A*, 1352–1361. [CrossRef]

14. Cao, W.; Wang, C.; Shi, J.; Wang, M.; Hui, W.; Dong, H. Microstructure and mechanical properties of Fe–0.2C–5Mn steel processed by ART-annealing. *Mater. Sci. Eng. A* **2011**, *528*, 6661–6666.

15. Sugimoto, K.; Tanino, H.; Kobayashi, J. Impact toughness of medium-Mn transformation-induced plasticity-aided steels. *Steel Res. Int.* **2015**, *86*, 1151–1160.

16. Tsuchiyama, T.; Inoue, T.; Tobata, J.; Akama, D.; Takaki, S. Microstructure and mechanical properties of a medium manganese steel treated with interrupted quenching and intercritical annealing. *Scr. Mater.* **2016**, *122*, 36–39. [CrossRef]

17. Available online: https://www.nissan-global.com/JP/TECHNOLOGY/OVERVIEW/high_ten.html (accessed on 29 September 2019).

18. Billur, E.; Cetin, B.; Gurleyik, M. New generation advanced high strength steels: Developments, trends and constraints. *Int. J. Sci. Technol. Res.* **2016**, *2*, 50–62.

19. Yoshioka, N.; Tachibana, M. Weight reduction of automotive seat components using high-strength steel with high formability. *Kobe Steel Eng. Rep.* **2017**, *66*, 22–25.

20. Wang, W.; Wang, H.; Liu, S. Development of hot-rolled Q&P steel at Baosteel. In Proceedings of the 9th Pacific Rim International Conference on Advanced Materials and Processing (PRICM9), Kyoto, Japan, 1–5 August 2016.

21. Wurm, T.; Busse, A.; Raedt, H. The lightweight forging initiative—Phase III: Material lightweight design for hybrid cars and heavy-duty trucks. *ATZ* **2019**, *121*, 16–21. [CrossRef]

22. Raedt, H.; Wurm, T.; Busse, A. The lightweight forging initiative—Phase III: Lightweight forging design for hybrid cars and heavy-duty trucks. *ATZ.* **2019**, *121*, 54–59. [CrossRef]

23. Yoshikawa, N.; Kobayashi, J.; Sugimoto, K. Notch-fatigue properties of advance TRIP-aided bainitic ferrite steels. *Metall. Mater. Trans. A* **2012**, *44A*, 4129–4136. [CrossRef]

24. Sugimoto, K. Fracture strength and toughness of ultrahigh strength TRIP-aided steels. *Mater. Sci. Technol.* **2009**, *25*, 1108–1117. [CrossRef]

25. Kobayashi, J.; Ina, D.; Nakajima, Y.; Sugimoto, K. Effects of microalloying on the impact toughness of ultrahigh-strength TRIP-aided martensitic steels. *Metall. Mater. Trans. A* **2013**, *44A*, 5006–5017. [CrossRef]

26. Mašek, B.; Jirková, H. New material concept for forging. In Proceedings of the International Conference on Advances in Civil, Structural and Mechanical Engineering (ACSME 2014), Bangkok, Thailand, 4–5 January 2014; pp. 92–96.

27. Bagliani, E.P.; Santofimia, M.J.; Zhao, L.; Sietsma, J.; Anelli, E. Microstructure, tensile and toughness properties after quenching and partitioning treatments of a medium-carbon steel. *Mater. Sci. Eng. A* **2013**, *559*, 486–495. [CrossRef]

28. Keul, C.; Wirths, V.; Bleck, W. New bainitic steel for forgings. *Arch. Civ. Mech. Eng.* **2012**, *12*, 119–125. [CrossRef]

29. Buchmayr, B. Critical assessment 22: Bainitic forging steels. *Mater. Sci. Technol.* **2016**, *32*, 517–522. [CrossRef]

30. Bleck, W.; Bambach, M.; Wirths, V.; Stieben, A. Microalloyed engineering steels with improved performance—An Overview. *Heat Treat. Mater. J.* **2017**, *72*, 355–364. [CrossRef]

31. Wirths, V.; Bleck, W. Carbide free bainitic forging steels with improved fatigue properties. In Proceedings of the International Conference on Steels in Cars and Trucks (SCT2014), Braunschweig, Germany, 15–19 June 2014; pp. 350–357.

32. De Diego-Calderón, I.; Sabirov, I.; Molina-Aldareguia, J.M.; Föjer, C.; Thiessen, R.; Petrov, R.H. Microstructural design in quenched and partitioned (Q&P) steels to improve their fracture properties. *Mater. Sci. Eng. A* **2016**, *657*, 136–146.

33. Gao, G.; An, B.; Zhang, H.; Guo, H.; Gui, X.; Bai, B. Concurrent enhancement of ductility and toughness in an ultrahigh strength lean alloy steel treated by bainite-based quenching-partitioning-tempering process. *Mater. Sci. Eng. A* **2017**, *702*, 104–122. [CrossRef]

34. Sugimoto, K.; Itoh, M.; Hojo, T.; Hashimoto, S.; Ikeda, S.; Arai, G. Microstructure and mechanical properties of ausformed ultra high-strength TRIP-aided steels. *Mater. Sci. Forum* **2007**, *539–543*, 4309–4314. [CrossRef]

35. Sugimoto, K.; Sato, S.; Arai, G. Hot forging of ultrahigh-strength TRIP-aided steel. *Mater. Sci. Forum* **2010**, *638–642*, 3074–3079. [CrossRef]

36. Sugimoto, K.; Kobayashi, J.; Arai, G. Development of ultrahigh-strength low alloy TRIP-aided steel for hot-forging parts. *Steel Res. Int. (Spec. Ed. Met. Form. 2010)* **2010**, *81*, 254–257.

37. Maruyama, H. X-ray measurement of retained austenite volume fraction. *J. Jpn. Soc. Heat Treat.* **1977**, *17*, 198–204.

38. Dyson, D.J.; Holmes, B.J. Effect of alloying additions on the lattice parameter of austenite. *Iron Steel Inst.* **1970**, *208*, 469–474.

39. Wu, J.; Wray, P.J.; Garcia, C.I.; Hua, M.; DeArdo, A.J. Image quality analysis: A new method of characterizing microstructures. *ISIJ Int.* **2005**, *45*, 254–262. [CrossRef]

40. Sugimoto, K.; Kobayashi, M.; Hashimoto, S. Ductility and strain-induced transformation in a high strength TRIP-aided dual-phase steel. *Metall. Trans. A* **1992**, *23A*, 3085–3091. [CrossRef]

41. Hojo, T.; Kobayashi, J.; Sugimoto, K. Impact properties of low-alloy transformation-induced plasticity-steels with different matrix. *Mater. Sci. Technol.* **2016**, *32*, 1–8. [CrossRef]

42. Rinebolt, J.A.; Harris, W.J., Jr. Effect of alloying elements on notch toughness of pearlitic steels. *Trans. Am. Soc. Met.* **1951**, *43*, 1175–1214.

43. Maisuradze, M.V.; Ryzhkov, M.A. Improving the impact toughness of the Hy-tuf steel by austempering. *AIP Conf. Proc.* **2018**, *2053*, 040054.

Review

An Overview of Fatigue Strength of Case-Hardening TRIP-Aided Martensitic Steels

Koh-ichi Sugimoto [1,*], Tomohiko Hojo [2] and Ashok Kumar Srivastava [3]

[1] Department of Mechanical Systems Engineering, School of Science and Technology, Shinshu University, Nagano 380-8553, Japan

[2] Institute for Materials Research, Tohoku University, Sendai 980-8577, Japan; hojo@imr.tohoku.ac.jp

[3] Department of Metallurgical and Materials Engineering, School of Engineering, OP Jindal University, Raigarh 496001, India; ashok.srivastava@opju.ac.in

* Correspondence: sugimot@shinshu-u.ac.jp; Tel.: +81-90-9667-4482

Received: 19 April 2018; Accepted: 5 May 2018; Published: 15 May 2018

Abstract: Surface-hardened layer characteristics and fatigue strength properties of transformation-induced plasticity-aided martensitic steels subjected to heat-treatment or vacuum carburization followed by fine-particle peening are revealed for automotive applications specially for powertrain parts. The as-heat-treated steels without the case-hardening process possess excellent impact toughness and fatigue strength. When the steels are subjected to fine-particle peening after heat-treatment, the fatigue limits of smooth and notched specimens increase considerably, accompanied with low notch sensitivity. Vacuum carburization and subsequent fine-particle peening increases further the fatigue strength of the steels, except notch fatigue limit. The increased fatigue limits are principally associated with high Vickers hardness and compressive residual stress just below the surface, resulting from the severe plastic deformation and the strain-induced martensitic transformation of metastable retained austenite, as well as low surface roughness and fatigue crack initiation depth.

Keywords: TRIP-aided martensitic steel; case-hardening; vacuum carburization; fine-particle peening; fatigue strength

---

1. Introduction

Recently, four types of transformation-induced plasticity (TRIP)-aided steels with different matrix structures and different chemical compositions, namely *"TRIP-aided bainitic ferrite (TBF) steel"* [1–10], *"TRIP-aided martensitic (TM) steel"* [11–21], *"quenching and partitioning (Q&P) steel"* [22–31], and *"medium Mn TRIP-aided steel"* [32–39] have been developed as third-generation advanced high-strength steels (AHSSs) for automotive sheet products. These steels possess high toughness, fatigue strength and delayed fracture strength, as well as excellent cold formability, especially in TM steels [7,11,16–20]. Therefore, these steels are expected to be also a suitable material for high torque powertrain parts which are produced by cold and hot forging [40], in the same way as *"nanostructured bainitic steel"* [41–48].

Many researchers found that shot peening considerably increases the fatigue strength of the conventional case-hardening martensitic steels [49–59]. Sugimoto et al. [60–62] have reported that high fatigue strength and low notch-sensitivity for fatigue of the TM steel are achieved by fine-particle peening and/or vacuum carburization because of small surface roughness and the improved surface-hardened layer properties. In addition, they showed that further increase in fatigue limit of the steels is achieved by using of multi process of vacuum carburization and fine-particle peening which increases the hardness and residual stress in the surface hardened layer [63–65].

In this paper, microstructural properties (matrix, second phase and retained austenite characteristics), mechanical properties (tensile, impact and fatigue properties) and relationships between the microstructural and mechanical properties of as-heat-treated (0.1–0.6)%C–1.5%Si–1.5%Mn–Cr–Mo–Ni TM steels are first introduced to understand the internal characteristics of case-hardening TM steel. Next, surface-hardened layer properties and fatigue strength of 0.2%C–1.5%Si–1.5%Mn–1.0%Cr–(0–0.2)%Mo TM steels subjected to heat-treatment or vacuum-carburization followed by fine-particle peening are explained. Besides, the fatigue strength properties are mainly correlated with the Vickers hardness and the residual stress in the surface-hardened layer which are enhanced by the strain-induced martensite transformation of metastable retained austenite and severe plastic deformations, as well as the surface roughness and fatigue crack initiation depth. This review paper is based on our researches referring to the fatigue strength of TM steels because no researcher could identify and address potential issues.

2. As-Heat-Treated TM Steels

2.1. Microstructure

C–Si–Mn–Cr–Mo–Ni TM steels with tensile strength higher than 1470 MPa can be produced by the multi-step heat-treatment consisting of austenitization and subsequent isothermal transformation at temperature below martensite-finish temperature (M_f) shown in Figure 1 [11–21]. On cooling to the isothermal transformation temperature, most of the austenite first transforms to soft α'-martensite with wide lath structure, with a small amount of untransformed austenite. On subsequent cooling after isothermal transformation, the austenite changes into a mixture of hard α'-martensite with narrow lath structure and carbon-enriched retained austenite (MA)-like phase. Resultantly, microstructure of the TM steels principally consists of soft α'-martensite matrix and MA-like phase (Figure 2) [13]. Most of the MA-like phases are located on prior austenitic grain, packet and block boundaries. The retained austenite phases are mainly filmy and in existence along narrow lath-martensite boundary in the MA-like phase, as shown in Figure 3. Fine and needle shaped carbides seem to precipitate only in wide lath-martensite structure. This means that the carbides are already precipitated during auto-tempering on quenching prior to low temperature isothermal transformation. It is noteworthy that the carbide fraction is much lower than that of the conventional low alloy martensitic steels with low content of Si.

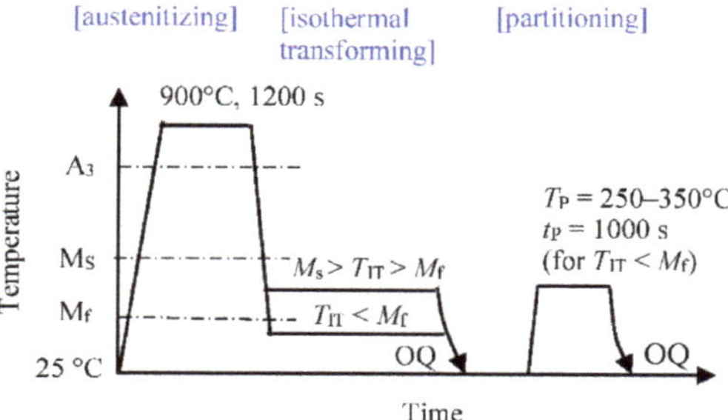

Figure 1. Multi-step heat-treatment diagram (austenitizing and subsequent isothermal transforming, followed by partitioning) of TM steel [16]. Isothermal transformation time is t_{IT} = 1000 s. Subscripts "IT" and "P" denote isothermal transforming and partitioning, respectively. OQ: quenching in oil at 50 °C.

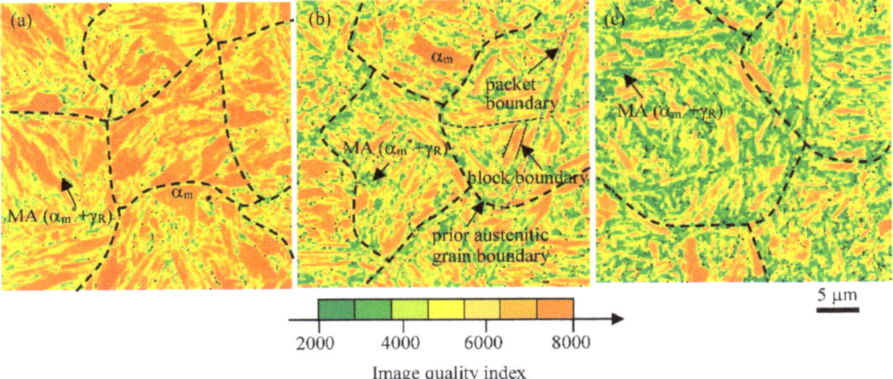

Figure 2. Phase maps of (**a**) 0.2%C–, (**b**) 0.4%C– and (**c**) 0.6%C–1.5%Si–1.5%Mn TM steels subjected to isothermal transforming at 50 °C for 1000 s after austenitizing [13]. Orange and yellowish green phases denote coarse lath-martensite (α_m) and MA-like phase (MA), respectively. Dark green dots in MA-like phase are retained austenite (γ_R). α_m* is narrow lath-martensite.

If isothermal transformation was conducted at low temperature such as room temperature, the retained austenite is not carbon-enriched sufficiently (is not stabilized mechanically). In this case, partitioning treatment at temperatures between M_S and M_f is usually employed after the isothermal transforming; the partitioning is different from the conventional low temperature tempering because it aims to carbon-enrichment into the retained austenite and soften further the soft α'-martensite, without dissolution of the retained austenite and carbide precipitation [11,12,15].

Volume fractions of retained austenite and MA-like phase increase with increasing carbon content of steel and isothermal transformation temperature at temperatures between M_S and M_f [16,19]. In this case, carbide fraction decreases with increasing isothermal transformation temperature. At near M_S, no carbide precipitates in the wide lath-martensite, in the same way as bainitic ferrite. The carbon concentration of retained austenite is independent of isothermal transformation temperature and nearly constant. Partitioning at temperatures lower than 350 °C for 1000 s increases carbon concentration of retained austenite although partitioning higher than 250 °C reduces only a little retained austenite fraction with an increase of carbide fraction [11,17].

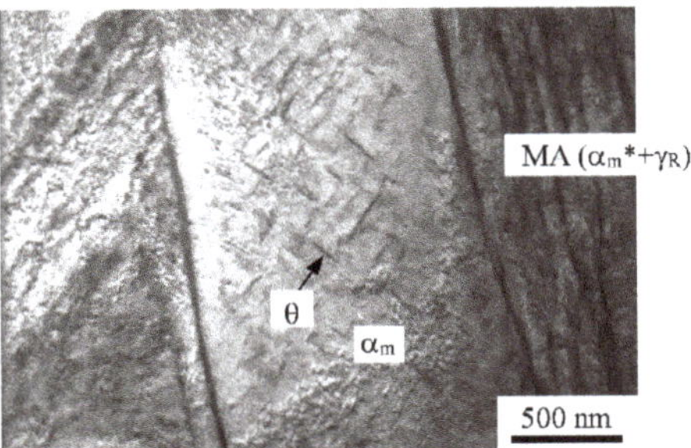

Figure 3. TEM image of 0.2%C–1.5%Si–1.5%Mn–1.0%Cr TM steel subjected to isothermal transforming at 50 °C for 1000 s [11]. α_m: wide lath-martensite, MA: a mixture of narrow lath-martensite (α_m*) and filmy retained austenite (γ_R), θ: carbide.

High carbon content increases the volume fractions of retained austenite and MA-like phase in (0.1–0.6)%C–1.5%Si–1.5%Mn TM steels with different carbon content (Figure 4). Higher carbon concentration of retained austenite can be obtained in the steels containing 0.2%C and 0.3%C [13,16]. The carbon concentration of retained austenite in 0.6%C steel is so much lower than the desired carbon content. This contradictory result may be due to large transformation strain of narrow lath-martensite. Microalloying of Cr, Mo, Ni and Mn also increases the volume fractions of retained austenite, MA-like phase and carbide, whereas decreases the carbon concentration of retained austenite [14,36].

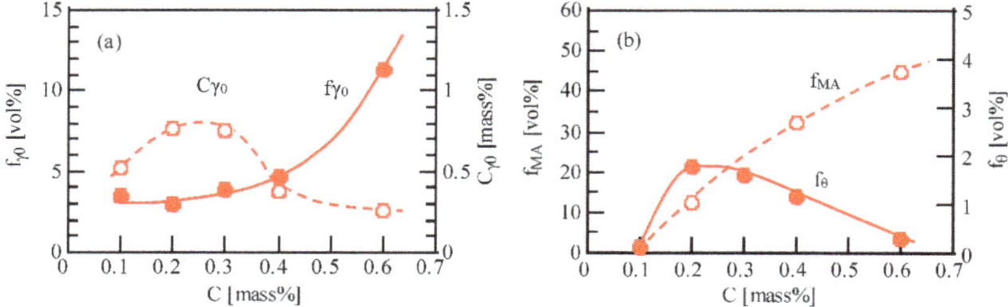

Figure 4. Variations in (**a**) initial volume fraction ($f\gamma_0$) and carbon concentration ($C\gamma_0$) of retained austenite and (**b**) volume fractions of MA-like phase (f_{MA}) and carbide (f_θ) as a function of carbon content (C) in (0.1–0.6)%C–1.5%Si–1.5%Mn TM steels subjected to isothermal transforming at 50 °C for 1000 s (without partitioning) [13].

Sometimes, the steel subjected to isothermal transforming at temperatures between martensite-start temperature (M_S) and M_f after austenitization is also regarded as TM steel because of similar microstructure [11] and mechanical properties [16], although the microstructure consists of soft α'-martensite/bainitic ferrite matrix and a larger amount of retained austenite than that of TM steel [11]. In this case, soft α'-martensite fraction can be estimated by using the empirical equation proposed by Koistinen and Marburger [66]. This multi-step heat-treatment resembles one-step Q&P process at temperatures between M_S and M_f.

2.2. Tensile Properties

When isothermal transformation treatment is carried out at temperatures below M_f in 0.2%C–1.5%Si–1.5%Mn–1.0%Cr–0.05%Nb TM steel, the tensile strength decreases and the total elongation increases with increasing isothermal transformation temperature [16]. In (0.1–0.6)%C–1.5%Si–1.5%Mn TM steels [13,16], the yield stress (or 0.2% offset proof stress) and tensile strength increase with increasing carbon content except for 0.6%C steel which fractures at an early stage (Figure 5). On the other hand, total and uniform elongations decrease with increasing carbon content. Resultantly, 0.4%C steel possesses the highest combination of tensile strength and total elongation (TS × TEl = 23 GPa %). The notch-tensile strength tends to increase with increasing carbon content in the same way as tensile strength. TM steels with carbon content of 0.2% to 0.4% exhibit higher notch-strength ratio than 0.1%C steel due to high TRIP effect of the retained austenite. The high notch-strength ratios are the same as those of quenched and tempered JIS-SCM420, 435 and 440 steels (Cr-Mo steels with different carbon content) [13]. Additions of Cr, Mo, Ni and Mn to 0.2%C–1.5%Si–1.5%Mn TM steel increase the yield stress and tensile strength, but decrease the elongations. Optimum combination of tensile properties can be obtained for 1.0%Cr bearing TM steel [14,17].

Sugimoto et al. [17] proposed that high combination of tensile strength and total elongation of TM steel is associated with the following characteristic microstructures;

(1)   a soft wide lath-martensite matrix with only a little carbide and diluted carbon concentration which improves the ductility,

(2)   a large quantity of finely dispersed MA-like phase which produces a large mean internal stress resulting from a difference in flow stress between a soft wide-lath martensite and a hard MA-like phase,

(3)   a metastable retained austenite of 2–5 vol %, which increases hard strain-induced martensite fraction and plastically relaxes the localized stress concentration on strain-induced transformation.

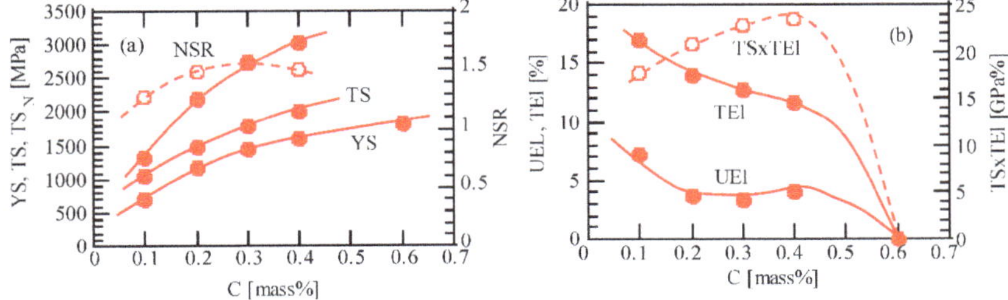

Figure 5. Variations in (**a**) yield stress or 0.2% offset proof stress (YS), tensile strength (TS), notch tensile strength and notch strength ratio (NSR = TS_N/TS) and (**b**) uniform and total elongations (UEL and TEL) and product of TS and TEl (TS × TEl) as a function of carbon content (C) in (0.1–0.6)%C–1.5%Si–1.5%Mn TM steels [13]. Stress concentration factor of notched specimen is 1.9.

2.3. Impact Toughness

It can be seen that when 1%Cr, 1%Cr–0.2%Mo, or 1%Cr–0.2%Mo–1.5%Ni is added to the base steel with a chemical composition of 0.2%C, 1.5%Si, 1.5%Mn and 0.05%Nb, the TM steels exhibit high V-notched Charpy impact absorbed values at 25°C ranging between 100 to 125 J/cm^2, and low ductile-brittle transition temperatures (corresponding to 50% ductile fracture appearance on Charpy impact tests) ranging from −150 to −130 °C (Figure 6) [14,19]. Moreover, the steels also possess the tensile strengths of around 1.4 to 1.5 GPa. The impact properties are far superior to those of the conventional low alloy martensitic steel (SCM 420 steel), in the same way as TBF and tempered Q&P [31] steels. In addition, the transition temperatures are considerably lowered as compared to TBF, tempered Q&P [31] and SCM420 steel. From the load–displacement curves obtained by impact testing, it is found that high impact values of TM steels are mainly caused by the high crack propagation energy [14]. Partitioning at 300 °C for 1000 s after isothermal transforming increases further impact values [12,14].

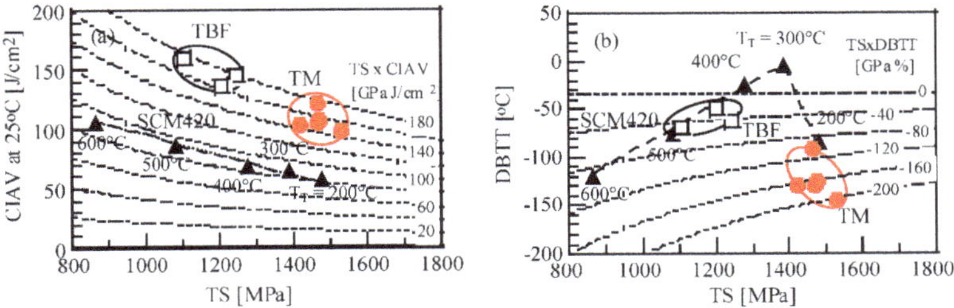

Figure 6.   Relationships between tensile strength (TS) and (**a**) Charpy impact absorbed value (CIAV) at 25°C and (**b**) ductile-brittle transition temperature (DBTT) for 0.2%C–1.5%Si–1.5%Mn–(0–1.0)%Cr–(0–0.2)%Mo–(0–1.5)%Ni–0.05%Nb TM steels (●) subjected to isothermal transforming at 200 °C, SCM420 steel tempered at T_T = 200 to 600 °C for 3600 s (▲) and TBF steels (□) [14]. The TBF steels have the same composition as TM steels and are subjected to isothermal transforming (or austempering) at 400 °C for 1000 s.

Figure 7 depicts the fracture mechanism of ductile and brittle impact fractures of TM steel [14]. In a case of ductile impact fracture, MA-like phases play an important role in suppressing the void initiation and promoting the preferential void growth at the MA-like phase/matrix interface (Figure 7a). In a case of brittle impact fracture, on the other hand, the MA-like phases also inhibit the initiation of cleavage cracking, as shown in Figure 7b, because they reduce the facet size of cleavage crack and the strain-induced transformation of the retained austenite relaxes plastically the localized stress concentration. With this, the superior impact properties of TM steel are believed to be caused by the characteristic microstructure of (1) to (3) mentioned in the last of Section 2.2.

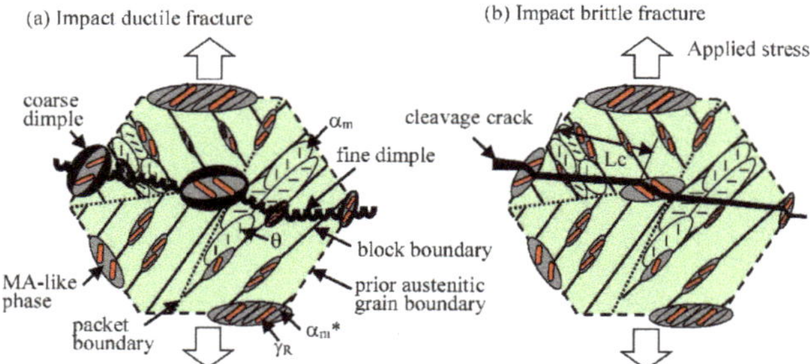

Figure 7. Illustrations showing (**a**) ductile and (**b**) brittle impact fractures for TM steel [14]. α_m, α_m*, γ_R, and θ represent wide lath-martensite, narrow lath-martensite, retained austenite, and carbide, respectively. *Lc*: quasi-cleavage crack path decided by the space between MA-like phases located on prior austenitic grain, packet and block boundaries.

2.4. Fatigue Properties

TM steel exhibits large fatigue hardening with subsequent softening in the same way as TBF steel (Figure 8a) [2,3,8,20], although commercial low alloy martensitic steel (SCM420 steel) softens during the cyclic deformation. Regardless of the high stress amplitude, a longer failure time and a larger plastic strain amplitude than those of SCM420 steel are also observed in the TM steel (Figure 8b), although the plastic strain is smaller than that of TBF steel. This cyclic hardening is mainly caused by "mean internal stress hardening" and "transformation hardening" [20] which result from the characteristic microstructures such as (1), (2) and (3) mentioned in the last of Section 2.2.

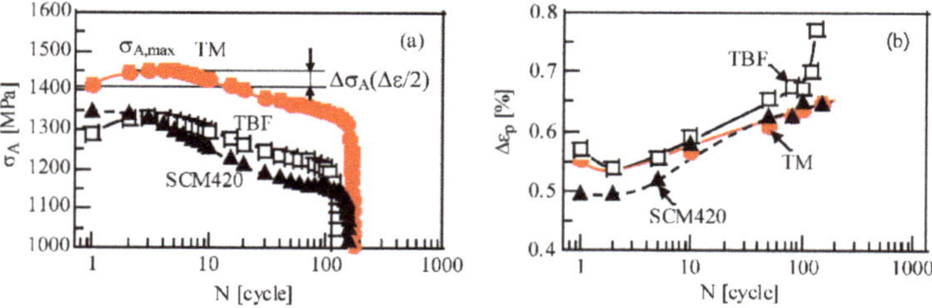

Figure 8. Variations in (**a**) stress amplitude (σ_A) and (**b**) plastic strain amplitude ($\Delta\varepsilon_p$) vs. the number of cycles (N) for 0.2%C–1.5%Si–1.5%Mn TM (●) and TBF (□) steels, and SCM420 steel (▲) [20]. Tests were carried out under tension-compression mode with a stress ratio of −1. Strain amplitude range is $\Delta\varepsilon$ = 1.5% and frequency is 1 Hz.

Figure 9 shows fatigue limits of smooth and notched specimens of (0.1–0.6)%C–1.5%Si–1.5%Mn TM steels with different carbon content [13,19] and the *"notch-sensitivity factor q"* [67] defined by the following equation,

$$q = (K_f - 1)/(K_t - 1) \tag{1}$$

where K_f and K_t are fatigue-notch factor ($= \sigma_w/\sigma_{wn}$; σ_w, σ_{wn}: fatigue limits of smooth and notched specimens, respectively) and stress concentration factor (1.9 in this study), respectively. Both fatigue limits of smooth and notched specimens increase with increasing carbon content in steels between 0.1%C and 0.4%C. The notch-sensitivity factor is the minimum in 0.3%C steel. Notched fatigue limits of the steels containing 0.2%C to 0.4%C are higher than those of SCM420, 435 and 440 steels with different carbon content. The notch-sensitivity factors of the TM steels are lower than those of SCM420, 435 and 440 steels. However, TBF steels possess relatively higher notch fatigue limits and far lower notch sensitivity factor than TM steels [7]. Similar tendency is also reported for fatigue limits in Q&P steels [24,27,28,41].

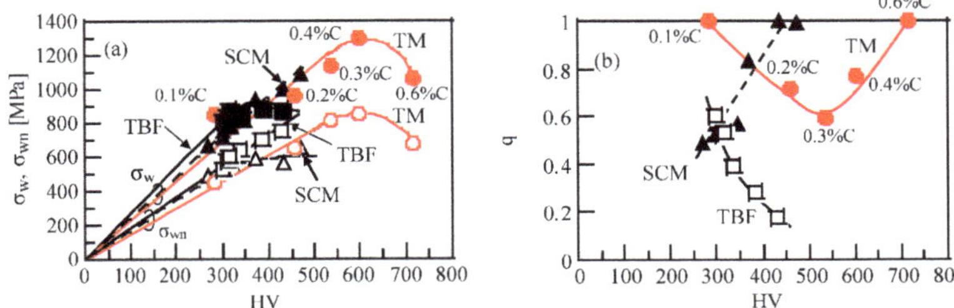

Figure 9. Variations in (**a**) fatigue limits of smooth and notched specimens (σ_w, σ_{wn}) and (**b**) notch sensitivity factor (q) as a function of Vickers hardness (HV) in (0.1–0.6)%C–1.5%Si–1.5%Mn TM steels ($\bigcirc\bullet$), 0.2%C–1.5%Si–1.5%Mn–(0–1.0)%Cr–(0–0.2)%Mo–(0–1.5)%Ni TBF steels ($\square\blacksquare$) and SCM420, 435 and 440 steels ($\triangle\blacktriangle$) [7,13]. Fatigue tests are conducted under tension-tension mode (a stress ratio of 0.1) at 25 °C with a sinusoidal wave of 80 Hz. The fatigue limits are decided by the maximum value of stress amplitude (a difference between maximum and minimum stresses) without failure up to 10^7 cycles.

In general, fatigue limits of smooth and notched specimens are principally controlled by fatigue crack initiation and propagation stages, respectively. Figure 10a shows a SEM image illustrating a fatigue crack initiated on the surface of a notched specimen of TM steel [13]. According to Knott [68], the plastic zone size d_Y of crack tip under plane strain condition (Figure 10b) can be estimated from the following equation:

$$d_Y = K^2/(6\pi YS^2) \tag{2}$$

where YS is the yield stress, and K is the stress intensity factor defined as $\sigma(\pi c)^{1/2}$, σ is the applied stress, and c is the crack length. From Figure 10a, it appears that the fatigue crack initiates in the wide lath martensite structure and is stopped at the MA-like phase [13]. In the 0.4%C steel, the plastic zone size is estimated to be about 2.0 µm if the fatigue crack length at the first stage is 2c = 30 µm, equivalent to the prior austenitic grain size, and the applied stress is corresponding to the fatigue limit. The plastic zone always includes some retained austenite particles in MA-like phase (see Figure 10b) because the interparticle path of MA-like phase is 0.5 to 2.0 µm (Figure 2b). Therefore, the plastic relaxation by the strain-induced transformation of retained austenite can be expected to take place always in the plastic zone during fatigue deformation. In conclusion, the high notch fatigue limits of TM steels can be considered to be principally caused by an increase in volume fraction of hard phase, plastic relaxation and crack closure via the strain-induced martensite transformation of metastable retained austenite, as well as a very small amount of carbide. Tomita et al. [41] also reports the similar improving mechanism for fatigue strength in 0.6%C–1.5%Si–0.8%Mn steel.

It is very difficult to compare the fatigue limits of TM steels with other AHSSs because fatigue tests are not conducted under the same fatigue mode and stress ratio. However, from comparison of several references [1,4,6,7,9,27,28,41,42,44–48], we can conclude that TM steels possess the high fatigue limits compared to the conventional quenched and tempered martensitic steels, in the same way as several AHSSs such as TBF steels [1,4,6,7,9], Q&P steels [27,28] and nanostructured bainitic steels [41,42,44–49].

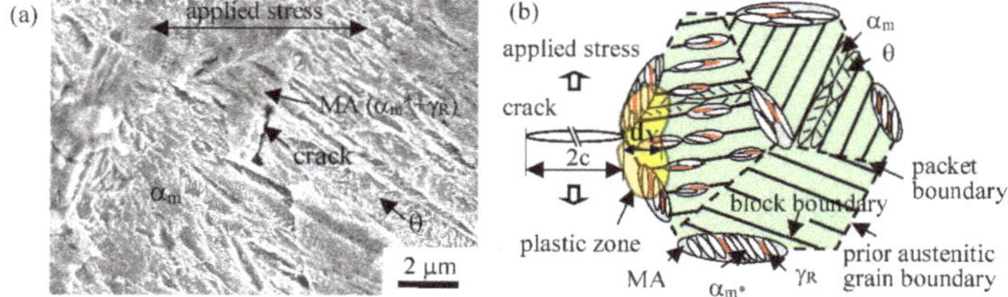

Figure 10. (**a**) SEM image of the initial crack formed on the surface of notched specimen in 0.4%C–1.5%Si–1.5%Mn TM steel, fractured at N_f = 5.0 × 10^4 cycles under tension-tension mode (a stress ratio of 0.1) at 25 °C with a sinusoidal wave of 80 Hz [13]. (**b**) An illustration of the plastic zone size (d_Y) in TM steel depicting the crack tip and distribution of MA-like phases: where α_m, α_m*, γ_R, θ and MA represent wide lath-martensite, narrow lath-martensite, retained austenite, carbide and MA-like phase, respectively.

3. Heat-Treated and Subsequently Fine-Particle Peened TM Steels

3.1. Surface-Hardened Layer Properties

When heat-treated 0.2%C–1.5%Si–1.5%Mn–1.0%Cr TM and JIS-SNCM420 steels (0.2%C–0.2%Si–0.5%Mn–0.6%Cr–0.2%Mo–1.7%Ni, quenched and tempered at 200 °C for 3600 s) are subjected to fine-particle peening under arc-height of 0.104 mm (N) at room temperature [62], the microstructures just below the surface are considerably refined and severely deformed without white layer (a nanosized or amorphous structure) [69], in the same way as TBF steel [60,61] (Figure 11b,d). In this case, arc-height of fine-particle peening is quantified using Almen strip of N type and the coverage is 300%. The maximum roughness height (*Rz*, JIS B0601:2001) on the surface of both steels is less than 4 μm because of the use of the small shot particles (steel ball of 70 μm diameter) (Figure 11a,c).

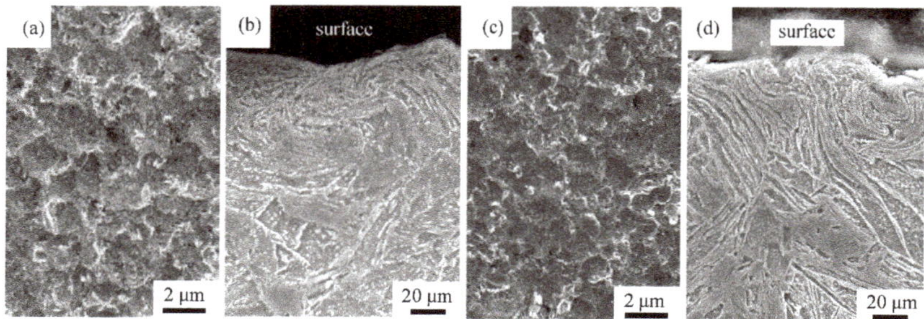

Figure 11. SEM images of (**a,c**) top-view of surface and (**b,d**) cross-sectional side view of surface-hardened layer of the smooth specimens of (**a,b**) heat-treated 0.2%C–1.5%Si–1.5%Mn–1.0%Cr TM and (**c,d**) SNCM420 steels subjected to fine-particle peening under arc-height of 0.104 mm (N) [62]. Maximum roughness heights of the TM and SNCM420 steels are *Rz* = 3.85 and 3.76 μm, respectively.

The distributions of Vickers hardness, volume fraction of untransformed retained austenite, and X-ray residual stress of the α(bcc) phase in the surface-hardened layer of the TM and SNCM420 steel specimens are shown in Figure 12 [62]. The depth of the surface-hardened layer before fatigue testing is approximately 20 μm in both steels; the maximum Vickers hardness increment (ΔHV, see Figure 12a) of the TM steel is larger than that of the SNCM420 steel. Most of the retained austenite transforms to martensite near the surface (Figure 12b). The ratio of the ΔHV to HV_0 (initial Vickers hardness, see Figure 12a) of TM steel increases with increasing volume fraction of strain-induced martensite ($\Delta f \alpha_m$), in the same manner as that of the TBF steel (Figure 13a) [61]. High compressive residual stresses occur in the α(bcc) phase of both steels before fatigue testing (Figure 12c). The compressive residual stress of the TM steel is larger than that of the SNCM420 steel. Maximum compressive residual stresses before fatigue are obtained at a depth of approximately 5–10 μm from the surface in both steels.

The maximum Vickers hardness and maximum compressive residual stress increase with increasing $\Delta f \alpha_m$ (Figure 13a,b). This indicates that the differences in Vickers hardness and compressive residual stress between both steels result from the differences in the $\Delta f \alpha_m$. The difference in the compressive residual stress is also associated with the expansion strain that occurs on the strain-induced martensite transformation.

After fatigue tests at the stress amplitude corresponding to fatigue limit, the surface-hardened layer is considerably softened in both steels (Figure 12a), with a slight decrease in the retained austenite fraction (Figure 12b). In particular, the Vickers hardness in the surface-hardened layer of SNCM420 steel is much lower than the original Vickers hardness. This behavior resembles cyclic softening of Figure 8a. It is noteworthy that the Vickers hardness of the TM steel increases even interior of the surface-hardened layer after fatigue deformation. Significant decreases in the compressive residual stress occur in both steels after fatigue test, although the compressive residual stress of the TM steel was slightly higher than that of SNCM420 steel (Figure 12c).

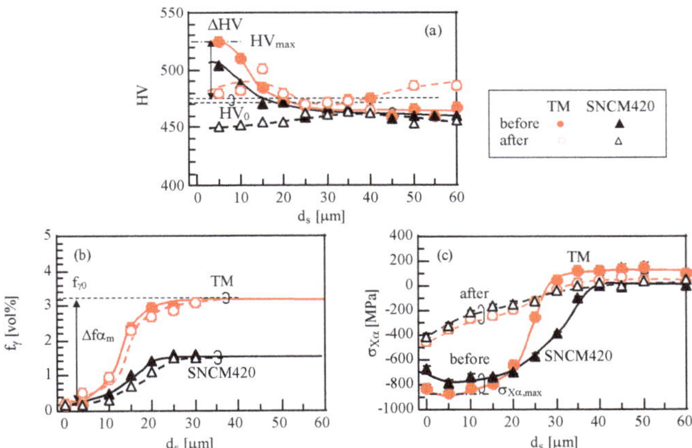

Figure 12. Distributions of (**a**) Vickers hardness (HV), (**b**) volume fraction of untransformed retained austenite ($f\gamma$) and (**c**) residual stress of the α(bcc) phase ($\sigma_{X\alpha}$) as a function of depth from the surface (ds) in heat-treated TM ($\bigcirc\bullet$) and SNCM420 ($\triangle\blacktriangle$) steels subjected to fine-particle peening under arc-height of 0.104 mm (N) before (solid marks) and after (open marks) fatigue testing [62]. Stress amplitudes (σ_a) of fatigue tests correspond to fatigue limits (TM: σ_a = 919 MPa; SNCM420: σ_a = 872 MPa).

3.2. Fatigue Strength

Heat-treated TM steel subjected to fine-particle peening under arc-height of 0.104 mm (N) achieves higher fatigue limits than SNCM420 steel, especially for the notched specimen (Figure 14). The TM steel also exhibits lower notch sensitivity factor q than the SNCM420 steel, although the q value for unpeened TM steel is higher than that for unpeened SNCM420 steel. It is considered that the fatigue limits of smooth and notched specimens in both steels are apparently enhanced by $\Delta f \alpha_m$, because the fatigue limits increase with increasing $\Delta f\alpha_m$ (Figure 13c) [62]. In Figure 13c, fatigue limits of SNCM420 steel tempered at 430 °C are added for the purpose of reference.

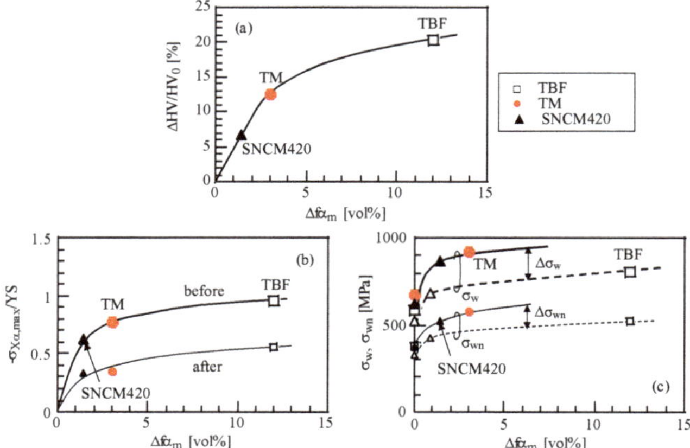

Figure 13. Variations in (**a**) hardness increment normalized to original hardness ($\Delta HV/HV_0$), (**b**) maximum compressive residual stress before and after fatigue normalized to yield stress ($-\sigma_{X\alpha,max}/YS$), and (**c**) fatigue limits of smooth and notched specimens (σ_w, σ_{wn}) as a function of the volume fraction of strain-induced transformed hard martensite ($\Delta f\alpha_m$) in 0.2%C–1.5%Si–1.5%Mn–1.0%Cr TM and TBF steels and SNCM420 steel subjected to fine-particle peening [62]. Fatigue limits of (**c**) were measured by rotary bending fatigue tests at 25 °C, using a sinusoidal wave of 3000 rpm and a stress ratio of $R = -1$. Marks $\triangle$ in (**c**) denote the fatigue limits of SNCM420 steel tempered at 430 °C.

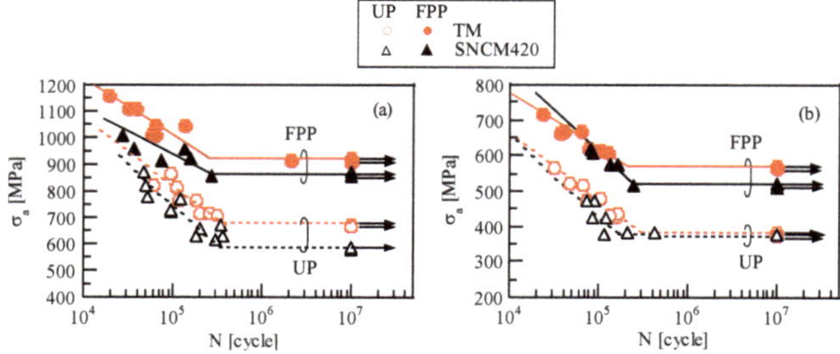

Figure 14. Stress amplitude–number of cycles (σ_a–N) curves for (**a**) smooth and (**b**) notched specimens in 0.2%C–1.5%Si–1.5%Mn–1.0%Cr TM and SNCM420 steels subjected to fine-particle peening (FPP) under arc-height of 0.104 mm (N) and also under unpeened (UP) conditions [62]. Fatigue tests were conducted on a rotating bending fatigue testing machine at 25 °C, using a sinusoidal wave of 3000 rpm and a stress ratio of $R = -1$.

Matsui et al. [55] reported that the sum of estimated yield stress ($\sigma_{Y,est}$) and the absolute value of maximum compressive residual stress of α(bcc) phase ($|\sigma_{X\alpha,max}|$) is related to the fatigue limit in JIS-SCM822H steel (0.22%C–0.27%Si–0.74%Mn–0.17%Cu–0.06%Ni–1.09%Cr–0.36%Mo) subjected to shot peening under arc-heights of 0.30 mm (N), 0.55 mm (A) and 0.25 mm (C) after gas-carburizing, as follows:

$$\sigma_w = 0.3891 \times (\sigma_{Y,est} + |\sigma_{X\alpha,max}|). \tag{3}$$

The value of $\sigma_{Y,est}$ can be estimated from HV_{max} by:

$$\sigma_{Y,est} = (HV_{max}/3) \times 9.80665 \times (YS/TS), \qquad (4)$$

where YS/TS is the yield ratio (assumed by Matsui et al. [55] to be 0.95).

The relationships between fatigue limits of smooth and notched specimens (σ_w and σ_{wn}) and $\sigma_{Y,est} + |\sigma_{X\alpha,max}|$ for the TM and SNCM420 steels are shown in Figure 15 [62]. The σ_ws of the TM and SNCM420 steels exhibit the same relationship as those by Matsui et al. [55], despite no carburizing. For a case of notched specimen of both steels, the following empirical equation is found:

$$\sigma_{wn} = 0.2282 \times (\sigma_{Y,est} + |\sigma_{X\alpha,max}|). \qquad (5)$$

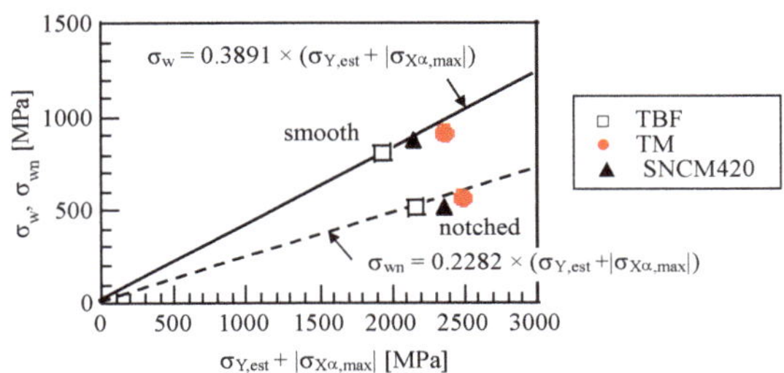

Figure 15. Relationship between the sum of estimated yield stress and maximum compressive residual stress ($\sigma_{Y,est} + \sigma_{X\alpha,max}$), and rotary bending fatigue limits (σ_w, σ_{wn}) for smooth and notched specimens of 0.2%C–1.5%Si–1.5%Mn–1.0%Cr TBF and TM steels and SNCM420 steel subjected to fine-particle peening under arc-height of 0.104 mm (N) [62].

4. Vacuum-Carburized and Subsequently Fine-Particle Peened TM Steels

4.1. Surface-Hardened Layer Properties

The surface hardened layer of 0.2%C–1.5%Si–1.5%Mn–1.0%Cr–0.2%Mo TM steel subjected to vacuum carburization under carbon potential of 0.8 mass% contains a large amount of retained austenite and a small amount of the MA-like phase, although coarsened microstructure appears near the surface (Figure 16a) [63]. The retained austenite fraction achieves to the maximum value (20.5 vol %) at a depth of 50 μm (Figure 17a). Two types of retained austenite, i.e., an island type in the matrix and a filmy type in the MA-like phase, appear in the steels (Figure 16a). A majority of the MA-like phase seems to be clustered along the prior austenitic, packet, and block boundaries. Vickers hardness of the as-vacuum-carburized TM steel is 713 at maximum (Figure 17c) and residual stress of α(bcc) phase is about −100 MPa (Figure 17e).

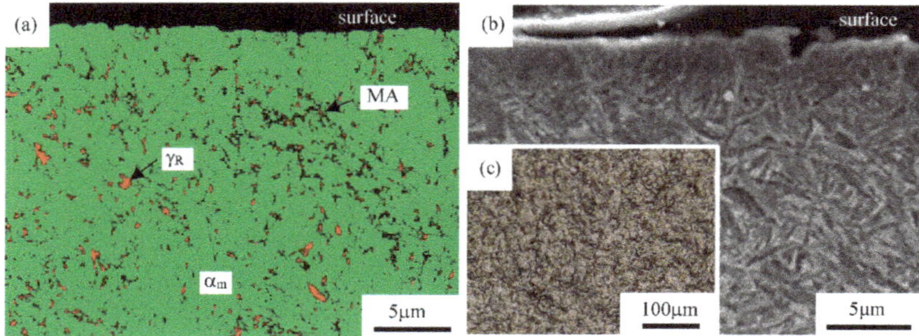

Figure 16. (**a**) EBSD phase map of 0.2%C–1.5%Si–1.5%Mn–1.0%Cr–0.2%Mo TM steel specimen subjected to vacuum carburization under carbon potential of 0.8 mass% and (**b**) SEM image of cross-sectional side view of surface-hardened layer and (**c**) optical image of surface of TM steel subjected to vacuum carburization and subsequent fine-particle peening under arc-height of 0.21 mm (N) by using steel ball of 50 μm [63,64]. Yellowish-green, red and gray regions represent α'-martensite (α_m) matrix, retained austenite (γ_R) phase and martensite-austenite (MA)-like phase, respectively. In (**c**), maximum roughness height of surface is Rz = 5.1 μm.

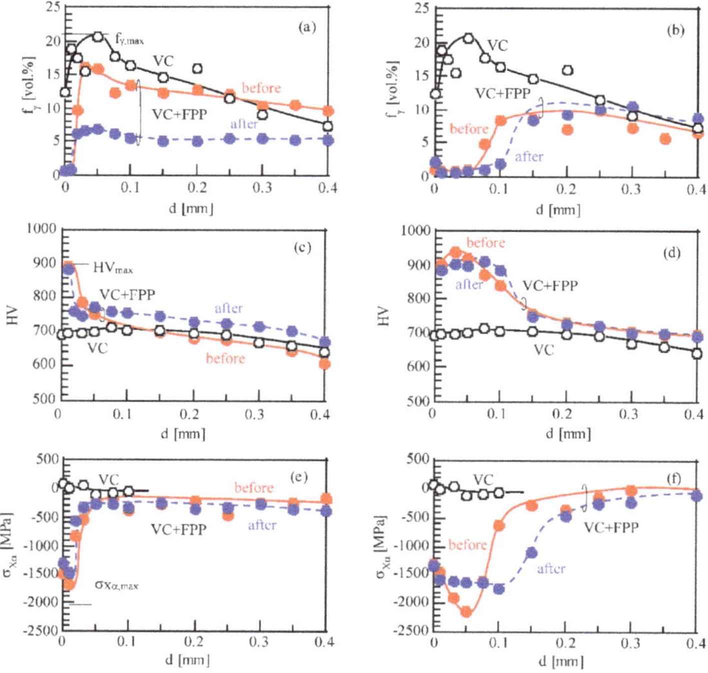

Figure 17. Distributions of (**a**,**b**) volume fractions of retained austenite (f_γ), (**c**,**d**) Vickers hardness (HV), and (**e**,**f**) residual stress of the α(bcc) phase ($\sigma_{X\alpha}$) before (●) and after (●) fatigue test in vacuum-carburized 0.2%C–1.5%Si–1.5%Mn–1.0%Cr–0.2%Mo TM steel subjected to vacuum-carburizing (VC) and then fine-particle peening (FPP) under arc-height of 0.21 mm (N) (**a**,**c**,**e**) and 1.12 mm (N) (**b**,**d**,**f**) [63,64]. Black open marks (○) and lines represent the surface-hardened layer properties of the as-vacuum-carburized steel. The *x*-axis shows the depth from the surface in mm.

After fine-particle peening, maximum roughness height of vacuum-carburized TM steel increases with incresing arc-height (Figure 18b), although the surface roughness is nearly the same under arc-heights lower than 0.21 mm (N). The typical surface image is shown in Figure 16c. White layer [69], which plays an important role in the fatigue strength and wear resistance is formed on the surface of the vacuum-carburized TM steel subjected to fine-particle peening under arc-heights higher than 0.41 mm (N). A distinct white layer is not formed on the surface in case of arc-heights lower than 0.21 mm (N) (Figure 16b). In the vacuum-carburized TM steel subjected to fine-particle peening under arc-height lower than 0.21 mm (N), the maximum retained austenite fraction is obtained at a depth of 50 μm (Figure 17a). The maximum Vickers hardness and the maximum compressive residual stress are attained at the depths of 10–50 μm. It is noteworthy that the Vickers hardness and compressive residual stress attain to the maximum values on surface or at near surface, as well as a large amount of untransformed retained austenite, when fine-particle peening is conducted under arc-height of 0.104 and 0.21 mm (N) (Figure 17c,e). The depths referring to their maximum values increase with increasing arc-height for the fine-particle peening (Figure 17b,d,f). In a case of higher arc-height, a larger amount of the retained austenite transforms to martensite (Figure 17b).

In vacuum-carburized TM and SNCM420 steels, the maximum volume fraction of retained austenite decreases with increasing arc-height (Figure 18a). On the other hand, the maximum Vickers hardness, the maximum residual stress and the maximum roughness height increase with increasing arc-height (Figure 18). The TM steel exhibits higher retained austenite fraction and compressive residual stress compared with SNCM420 steel, although the maximum Vickers hardness of both steels is almost the same. According to Sugimoto et al. [63], the maximum Vickers hardness and the maximum residual stress are mainly developed by severe plastic deformation, with a small contribution of the strain-induced martensitic transformation of the retained austenite.

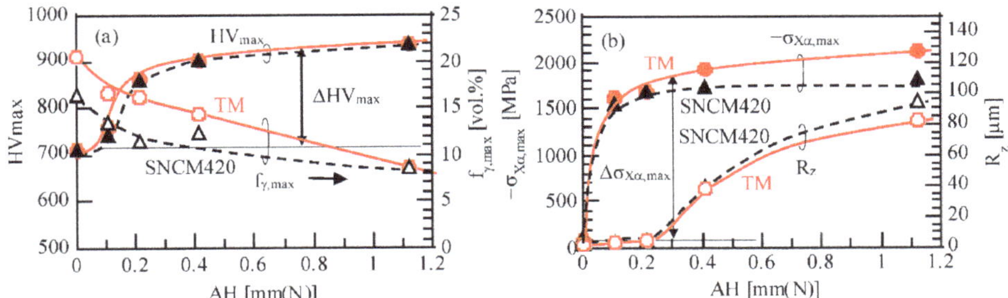

Figure 18. Variations in (**a**) the maximum retained austenite fraction ($f_{\gamma,\max}$) and the maximum Vickers hardness (HV$_{\max}$) and (**b**) the maximum compressive residual stress ($\sigma_{X\alpha,\max}$) and maximum roughness height (R_z) as a function of arc-height (AH) in TM ($\bullet\circ$) and SNCM420 ($\blacktriangle\triangle$) steels subjected to vacuum-carburizing and subsequent fine-particle peening [63].

4.2. Fatigue Properties

When vacuum-carburized TM and SNCM420 steels were subjected to fine-particle peening under arc-height of 0.21 mm (N), the fatigue limits of both steels exhibit the maximum values in an arc-height range of 0.104 to 1.22 mm (N) (Figure 19a) [64]. Moreover, the values of fatigue limit of both steels are similar, except for the fatigue limits of the vacuum-carburized steels. Kato et al. [53] also reported that the same arc-height achieved the highest fatigue limit in gas-carburized SCM822H steel (0.20%C–0.20%Si–0.75%Mn–0.15%Cu–0.10%Ni–1.10%Cr–0.15%Mo). The maximum increment in the fatigue limit ($\Delta\sigma_w$ in Figure 19a) and the percentage increase of vacuum-carburized TM steel due to fine-particle peening are 574 MPa and 82% at maximum, respectively. By the comparison of the fatigue limits with those [62] of heat-treated TM steels subjected to fine-particle peening under arc-height of 0.104 mm (N), the increase in the fatigue limit ($\Delta\sigma_w$ = 440 MPa) is greater than that ($\Delta\sigma_w$ = 244 MPa) of heat-treated steels (Figure19a). The increase in the fatigue limit is greater than those of gas-carburized martensitic steel subjected to shot peening or fine particle peening [57]. The optimum arc-height value for fatigue limit may vary if fine-particle peening is conducted under different shot material, angle of attack, velocity, etc. According to Sugimoto et al. [65], vacuum carburization followed by fine-particle peening increases notch sensitivity factor for fatigue because only few increase in the notch fatigue limit is obtained compared to that of smooth specimen.

During fatigue deformation, a large amount of untransformed retained austenite transforms to martensite in vacuum-carburized TM steel subjected to fine-particle peening under low arc-height (Figure 17a). According to study [62] using heat-treated TM steel (containing retained austenite fraction of 3.1 vol %) subjected to fine-particle peening under the arc-height of 0.104 mm (N), the Vickers hardness and the residual stress in the surface-hardened layer considerably decreased during a fatigue test at a stress amplitude corresponding to the fatigue limit. In this study, however, their declines are relatively slight and the Vickers hardness at a depth of more than 50 μm increases during fatigue deformation in vacuum-carburized TM steel subjected to fine particle peening under arc-height of 0.21 mm (N) (Figure 17c). The strain-induced martensite fraction during fatigue deformation of vacuum-carburized and fine-particle peened TM steel is far higher than that of heat-treated TM steel subjected to fine particle peening (Figure 17a). Therefore, this result indicates that the strain-induced martensite transformation of a large amount of untransformed retained austenite during fatigue deformation compensates large decreases in Vickers hardness and compressive residual stress and highly maintains the properties during fatigue deformation. Resultantly, the small declines of the Vickers hardness and compressive residual stress during fatigue deformation is considered to increase considerably the fatigue limit.

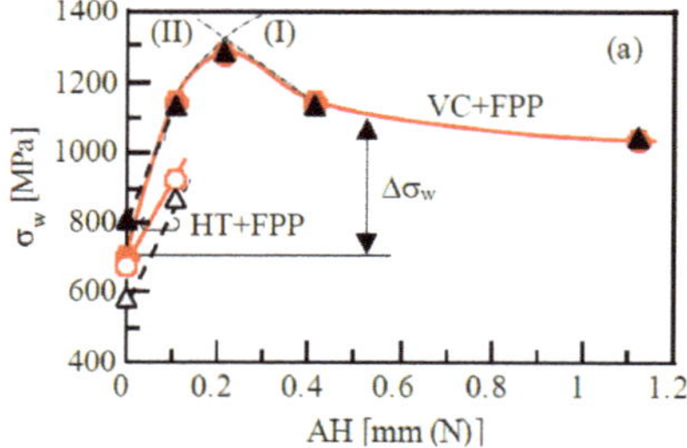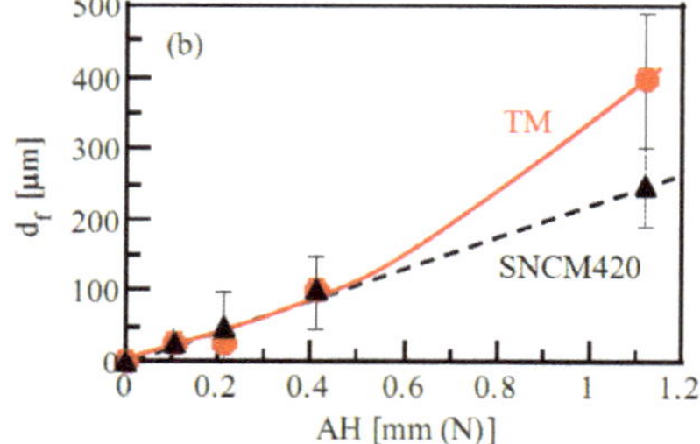

Figure 19. Variations in (**a**) the fatigue limit (σ_w) and (**b**) the depth of crack origin from the surface (d_f) as a function of the arc-height (AH) in TM (●○) and SNCM420 (▲△) steels subjected to vacuum-carburizing and fine-particle peening (VC+FPP) [64]. In (**a**), marks ○ and △ represent fatigue limits of TM (original retained austenite fraction $f\gamma_0$ = 3.2 vol %) and SNCM420 ($f\gamma_0$ = 1.6 vol %) steels subjected to heat-treatment and fine-particle peening (HT+FPP), respectively. (I): increases in Vickers hardness and compressive residual stress, (II): fish-eye crack fracture at a high depth and very few strain-induced martensite transformation. Fatigue tests were conducted on a rotating bending fatigue testing machine at 25 °C, using a sinusoidal wave of 3000 rpm and a stress ratio of $R = -1$.

In Figure 20, the fatigue limits of vacuum-carburized TM steel subjected to fine-particle peening under the arc-height of 0.104 and 0.21 mm (N) reduce by decline (ii), compared to those of heat-treated TM steel although they linearly increase with increasing $\sigma_{Y,est} + |\sigma_{X\alpha,max}|$ (line (i)). Meanwhile, the fatigue limits under arc-height of 0.41 and 1.12 mm (N) shift down by decline (iii), despite higher maximum Vickers hardness and maximum compressive residual stress in the surface-hardened layer than those under lower arc-height. Sugimoto et al. [64] explain the declines (ii) and (iii) as follows.

The declines in fatigue limits of vacuum-carburized TM steel are related to the depth of fish-eye crack from surface because most of the fatigue failure occurrs by fish-eye cracks (Figure 21), different from that of heat-treated TM steel [62]. The fish-eye crack depth increases with increasing arc-height of fine-particle peening (Figures 19b and 21). Therefore, decline (ii) in fatigue limit is considered to be caused by fish eye fracture. Decline (iii) in fatigue limits is associated with deep fish-eye fracture and a small amount of strain-induced martensite transformation during fatigue deformation (Figure 17b). In this case, superior surface-hardened layer properties and a white layer formed by fine particle peening under the arc-height higher than 0.41 mm (N) hardly contribute to increase the fatigue limit (Figure 21b), although a little strain-induced martensite plays a role in slight plastic relaxation of the localized stress concentration [70].

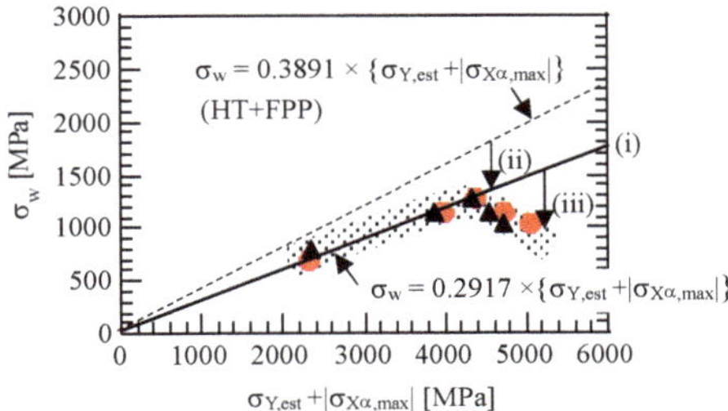

Figure 20. Relationship between the rotary bending fatigue limit (σ_w) and the sum of estimated yield stress and absolute value of maximum compressive residual stress ($\sigma_{Y,est} + |\sigma_{X\alpha,max}|$) of TM (●) and SNCM420 (▲) steels subjected to vacuum-carburizing and subsequent fine-particle peening [64]. (i): σ_w of vacuum-carburized and subsequently fine-particle peened steels, (ii): fish-eye crack fracture at a low depth, (iii): fish-eye crack fracture at a high depth and very few strain-induced transformation during fatigue deformation. HT+FPP: heat-treatment plus subsequent fine particle peening.

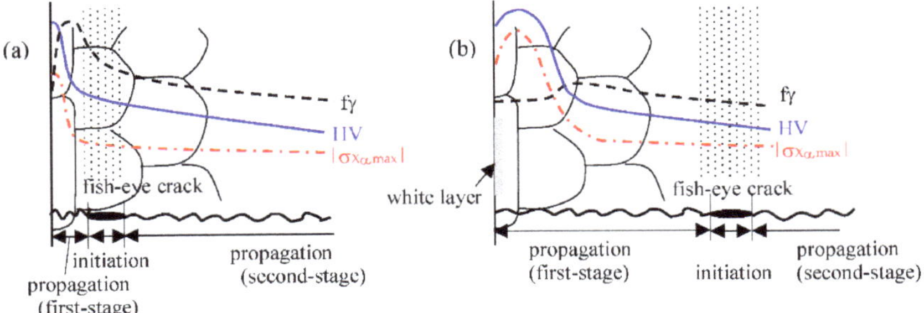

Figure 21. Distributions of surface-hardened layer properties and illustrations of initiation and propagation stages of fish-eye cracks in the vacuum-carburized TM steel subjected to fine-particle peening under arc-heights of (**a**) 0.104 and 0.21 mm (N) and (**b**) 0.41 and 1.12 mm (N) [64]. $f\gamma$: volume fraction of retained austenite; HV: Vickers hardness; $|\sigma_{X\alpha,max}|$: absolute value of residual stress in the α(bcc) phase.

5. Conclusions

The as-heat-treated TM steels possessed excellent impact toughness (high Charpy impact absorbed value and low ductile-brittle transition temperature) and high fatigue strength (large cyclic softening, high fatigue limit and low notch-sensitivity). When the steels are subjected to fine-particle peening after heat-treatment, the fatigue limits of smooth and notched specimens increased considerably, accompanied with low notch sensitivity. Vacuum carburization and subsequent fine-particle peening increased further the fatigue strength of the steels, except notch fatigue limit. The increased fatigue limits were principally associated with high Vickers hardness and compressive residual stress just below the surface, resulting from the severe plastic deformation and the strain-induced martensitic transformation of metastable retained austenite, as well as low surface roughness and fatigue crack initiation depth.

Further increases in the mechanical properties are possible by hot-forging prior to case-hardening, which brings on an increase in retained austenite fraction and a refined microstructure. Resultantly, the hot-forging is expected to improve the toughness and fatigue strength [12,15,18], in the same way as TBF steels [71–73]. The use of C, Cr and Mo bearing TM steels makes easy to manufacture the hot-forged parts as drivetrains, Diesel engine common rail systems, and others [40] and resultantly achieve the lightweight and low fuel consumption, instead of conventional high-strength low alloy steels. It is greatly hoped that many researchers interest in the hot-forging of TM steel and investigate the mechanical properties, such as fatigue strength, toughness and wear property of the case-hardening hot-forging products.

Acknowledgments: This study was supported by a Grant-in-Aid for Scientific Research (B), the Ministry of Education, Science, Sports and Culture, Japan (No. 2013-25289262).

Conflicts of Interest: The authors declare no conflict of interest.

References

1. Sugimoto, K.; Kobayashi, M.; Inoue, K.; Sun, X.; Soshiroda, T. Fatigue strength of TRIP-aided bainitic sheet steels. *Tetsu-to-Hagane* **1998**, *52*, 559–565. [CrossRef]
2. Sugimoto, K.; Kobayashi, M.; Inoue, K.; Masuda, S. Fatigue-hardening behavior of TRIP-aided bainitic steels. *Tetsu-to-Hagane* **1999**, *85*, 856–862. [CrossRef]
3. Sugimoto, K.; Song, S.; Inoue, K.; Kobayashi, M.; Masuda, S. Effect of prestraining on low cycle fatigue properties of low alloy TRIP steels. *J. Soc. Mater. Sci. Jpn.* **2001**, *50*, 657–664. [CrossRef]
4. Song, S.; Sugimoto, K.; Kandaka, S.; Futamura, A.; Kobayashi, M.; Masuda, S. Effects of prestraining on high cycle fatigue strength of high strength low alloy TRIP steels. *J. Soc. Mater. Sci. Jpn.* **2001**, *50*, 1091–1097. [CrossRef]
5. Sugimoto, K. Fracture strength and toughness of ultrahigh-strength TRIP aided steels. *Mater. Sci. Technol.* **2009**, *25*, 1108–1117. [CrossRef]
6. Sugimoto, K.; Yoshikawa, N. Advanced high-strength TRIP-aided steels for ultra high pressure DI-diesel engine common rail. In Proceedings of the 3rd International Conference on Stainless Steel in Cars and Trucks (SCT2011), Stuttgart, Germany, 5–9 June 2011; pp. 365–372.
7. Yoshikawa, N.; Kobayashi, J.; Sugimoto, K. Notch fatigue properties of advanced TRIP-aided bainitic ferrite steels. *Metall. Mater. Trans. A* **2012**, *43*, 4129–4136. [CrossRef]
8. Zhou, Q.; Qian, L.; Meng, J.; Zhao, L.; Zhang, F. Low-cycle fatigue behavior and microstructure evolution in a low carbon carbide-free bainitic steel. *Mater. Des.* **2015**, *85*, 487–496. [CrossRef]
9. El-Din, H.N.; Showaib, E.A.; Zaafarani, N.; Refaiy, H. Structure-properties relationship in TRIP type bainitic ferrite steel austempered at different temperatures. *Int. J. Mech. Mater. Eng.* **2017**, *12*, 3. [CrossRef]
10. Sugimoto, K.; Hojo, T.; Kobayashi, J. Critical assessment of TRIP-aided bainitic ferrite steels. *Mater. Sci. Technol.* **2017**, *33*, 2005–2007. [CrossRef]
11. Kobayashi, J.; Song, S.; Sugimoto, K. Microstructure and retained austenite characteristics of ultrahigh-strength TRIP-aided martensitic steels. *ISIJ Int.* **2012**, *52*, 1124–1129. [CrossRef]
12. Kobayashi, J.; Sugimoto, K.; Arai, G. Effects of hot-forging process on combination of strength and toughness in ultrahigh-strength TRIP-aided martensitic steels. *Adv. Mater. Res.* **2012**, *409*, 696–701. [CrossRef]
13. Kobayashi, J.; Yoshikawa, N.; Sugimoto, K. Notch-fatigue strength of advanced TRIP-aided martensitic steels. *ISIJ Int.* **2013**, *53*, 1479–1486. [CrossRef]
14. Kobayashi, J.; Ina, D.; Nakajima, Y.; Sugimot, K. Effects of microalloying on the impact toughness of ultrahigh-strength TRIP-aided martensitic steels. *Metall. Mater. Trans. A* **2013**, *44*, 5006–5017. [CrossRef]
15. Sugimoto, K.; Kobayashi, J.; Hojo, T. Mechanical properties of TRIP-aided martensitic steels for hot-worked automotive drivetrain components. In Proceedings of the 4th International Conference on Steels in Cars and Trucks (SCT), Braunschweig, Germany, 15–19 June 2014; pp. 21–28.

16. Kobayashi, J.; Ina, D.; Futamura, A.; Sugimoto, K. Fracture toughness of an advanced ultrahigh-strength TRIP-aided steels. *ISIJ Int.* **2014**, *54*, 955–962. [CrossRef]

17. Pham, D.V.; Kobayashi, J.; Sugimoto, K. Effects of microalloying on stretch-flangeability of ultrahigh-strength TRIP-aided martensitic steel sheet. *ISIJ Int.* **2014**, *54*, 1943–1951. [CrossRef]

18. Hojo, T.; Kobayashi, J.; Kochi, T.; Sugimoto, K. Effects of thermomechanical processing on microstructure and shear properties of 22SiMnCrMoB TRIP-aided martensitic steel. *Iron Steel Technol. Mag.* **2015**, *12*, 102–110.

19. Sugimoto, K.; Srivastava, A.K. Microstructure and mechanical properties of a TRIP-aided martensitic steel. *Metall. Microstr. Anal.* **2015**, *4*, 344–354. [CrossRef]

20. Sugimoto, K.; Hojo, T. Fatigue hardening behavior of a 1.5 GPa Grade TRIP-aided martensitic steel. *Metall. Mater. Trans. A* **2016**, *47*, 5272–5279. [CrossRef]

21. Zhao, P.; Cheng, C.; Gao, G.; Hui, W.; Misra, R.D.K.; Bai, B.; Weng, Y. The potential significance of microalloying with niobium in governing very high cycle fatigue behavior of bainit/martensite multiphase steels. *Mater. Sci. Eng. A* **2016**, *650*, 438–444. [CrossRef]

22. Speer, J.G.; Matlock, D.K.; De Cooman, B.C.; Schroth, J.G. Carbon partitioning into austenite after martensite transformation. *Acta Mater.* **2003**, *51*, 2611–2622. [CrossRef]

23. Speer, J.G.; De Moor, E.; Findley, K.O.; Matlock, D.K.; De Cooman, B.C.; Edmonds, D.V. Analysis of microstructure evolution in quenching and partitioning automotive sheet steel. *Metall. Mater. Trans. A* **2011**, *42*, 3591–3601. [CrossRef]

24. Cerny, I.; Mikulova, D.; Sis, J.; Masek, B.; Jirkova, H.; Malina, J. Fatigue properties of a low alloy 42SiCr steel heat treated by quenching and partitioning process. *Proc. Eng.* **2011**, *10*, 3310–3315. [CrossRef]

25. Toji, Y.; Matsuda, H.; Herbig, M.; Choi, P.; Raabe, D. Atomic-scale analysis of carbon partitioning between martensite and austenite by atom probe tomography and correlative transmission electron microscopy. *Acta Mater.* **2014**, *65*, 215–228. [CrossRef]

26. Wu, R.; Li, W.; Zhou, S.; Zhong, S.; Wang, L.; Jin, X. Effect of retained austenite on the fracture toughness of quenching and partitioning (Q&P)-treated sheet steels. *Metall. Mater. Trans. A* **2014**, *45*, 1892–1902.

27. De Diego-Calderon, I.; Rodriguez-Calvillo, A.; Lara, A.; Molina-Aldareguia, J.M.; Petrov, R.H.; De Knijf, D.; Sabirov, I. Effect of microstructure on fatigue behavior of advanced high strength steels produced by quenching and partitioning and the role of retained austenite. *Mater. Sci. Eng. A* **2015**, *641*, 215–224. [CrossRef]

28. Zhao, P.; Zhang, B.; Cheng, C.; Misra, R.D.K.; Gao, G.; Bai, B.; Weng, Y. The significance of ultrafine film-like retained austenite in governing very high cycle fatigue behavior in an ultrahigh-strength MN–SI–Cr–C steel. *Mater. Sci. Eng. A* **2015**, *645*, 116–121. [CrossRef]

29. Speer, J.G.; De Moor, E.; Clarke, A.J. Critical assessment 7: Quenching and partitioning. *Mater. Sci. Technol.* **2015**, *31*, 3–9. [CrossRef]

30. Seo, E.J.; Cho, L.; Estrin, Y.; De Cooman, B.C. Microstructure-mechanical properties relationships for quenching and partitioning (Q&P) processed steel. *Acta Mater.* **2016**, *113*, 124–139.

31. Gao, G.; An, B.; Zhang, H.; Guo, H.; Gui, X.; Bai, B. Concurrent enhancement of ductility and toughness in an ultrahigh strength lean alloy steel treated by bainite-based quenching-partitioning-tempering process. *Mater. Sci. Eng. A* **2017**, *702*, 104–112. [CrossRef]

32. Miller, R.L. Ultrafine-grained microstructures and mechanical properties of alloy steels. *Metall. Trans.* **1972**, *3*, 905–912. [CrossRef]

33. Furukawa, T. Dependence of strength-ductility characteristics on thermal history in low carbon 5 wt-%Mn steels. *Mater. Sci. Technol.* **1989**, *5*, 465–470. [CrossRef]

34. Furukawa, T.; Huang, H.; Matsumura, O. Effects of carbon content on mechanical properties of 5% Mn steels exhibiting transformation induced plasticity. *Mater. Sci. Technol.* **1994**, *10*, 964–969. [CrossRef]

35. Nakada, N.; Tsuchiyama, T.; Takaki, S.; Miyano, N. Temperature dependence of austenite nucleation behavior from lath martensite. *ISIJ Int.* **2011**, *51*, 299–304. [CrossRef]

36. Sugimoto, K.; Tanino, H.; Kobayashi, J. Impact toughness of medium Mn transformation-induced plasticity-aided steels. *Steel Res. Int.* **2015**, *86*, 1151–1160. [CrossRef]

37. Rana, R.; Gibbs, P.J.; De Moor, E.; Speer, J.G.; Matlock, D.K. A composite modeling analysis of the deformation behavior of medium manganese steels. *Steel Res. Int.* **2015**, *86*, 1139–1150. [CrossRef]

38. Tsuchiyama, T.; Inoue, T.; Tobata, J.; Akama, D.; Takaki, S. Microstructure and mechanical properties of a medium manganese steel treated with interrupted quenching and intercritical annealing. *Scr. Mater.* **2016**, *122*, 36–39. [CrossRef]

39. Qi, X.Y.; Du, L.X.; Hu, J.; Misra, R.D.K. High-cycle fatigue behavior of low-C medium-Mn high strength steel with austenite-martensite submicron-sized lath-like structure. *Mater. Sci. Eng. A* **2018**, *718*, 477–482. [CrossRef]

40. Raedt, H.W.; Wilke, F.; Ernst, C.S. Light weight forging initiative phase II: Lightweight design potential for a light commercial vehicle. *ATZ* **2015**, *118*, 48–52.

41. Tomita, Y.; Kijima, F.; Morioka, K. Modified austempering effect on bending fatigue properties of Fe-0.6C-1.5Si-0.8Mn steel. *Z. Metallkd.* **2000**, *91*, 43–46.

42. Bhadeshia, H.K.D.H. Nanostructure bainite. *Proc. R. Soc. Lond. A* **2010**, *466*, 3–18. [CrossRef]

43. Yang, J.; Wang, T.S.; Zhang, B.; Zhang, F.C. High-cycle bending fatigue behavior of nanostructured bainitic steel. *Scr. Mater.* **2012**, *66*, 363–366. [CrossRef]

44. Sourmail, T.; Caballero, F.G.; Garcia-Mateo, C.; Smanio, V.; Ziegler, C.; Kunts, M. Evaluation of potential of high Si high C steel nanostructured bainite for wear and fatigue applications. *Mater. Sci. Technol.* **2013**, *29*, 1166–1173. [CrossRef]

45. Shendy, B.R.; Yoozbashi, M.N.; Avishan, B.; Yazdani, S. An investigation on rotating bending fatigue behavior of nanostructured low-temperature bainitic steel. *Acta Metall. Sin.* **2014**, *27*, 233–238. [CrossRef]

46. Rementeria, R.; Morales-Rivas, L.; Kuntz, M.; Garcia-Mateo, C.; Kerscher, E.; Sourmail, T.; Caballero, F.G. On the role of microstructure in governing the fatigue behavior of nanostructured banitic steels. *Mater. Sci. Eng. A* **2015**, *630*, 71–77. [CrossRef]

47. Müller, I.; Rementeria, R.; Caballero, F.G.; Kuntz, M.; Sourmail, T.; Kerscher, E. A constitutive relationship between fatigue limit and microstructure in nanostructured bainitic steels. *Materials* **2016**, *9*, 831–849. [CrossRef] [PubMed]

48. Müller, I.; Rementeria, R.; Caballero, F.G.; Kuntz, M.; Kerscher, E. Correlation of fatigue limit and crack growth threshold value to the nanobainite microstructure. *Solid State Phenom.* **2016**, *258*, 314–317. [CrossRef]

49. Egami, N.; Kagaya, C.; Inoue, N.; Takeshita, H.; Mizutani, H. Hybrid surface modification of SCM415 material by vacuum carburizing and fine particle peening. *Jpn. Soc. Mech. Eng. A* **2000**, *66*, 1936–1942. [CrossRef]

50. Torres, M.A.S.; Voorwald, H.J.C. An evaluation of shot peening, residual stress and stress relaxation on the fatigue life of AISI 4340 steel. *Int. J. Fatigue* **2002**, *24*, 877–886. [CrossRef]

51. Shaw, B.A.; Aylott, C.; O'Hara, P.; Brimble, K. The role of residual stress on the fatigue strength of high performance gearing. *Int. J. Fatigue* **2003**, *25*, 1279–1283. [CrossRef]

52. Eto, H.; Matsui, K.; Jin, Y.; Ando, K. Influence of retained austenite, strain-induced martensite and pre-loaded stress upon compressive residual stress with shot peening method. *Jpn. Soc. Mech. Eng. A* **2003**, *69*, 733–740. [CrossRef]

53. Kato, M.; Matsumura, Y.; Ishikawa, R.; Kobayashi, Y.; Ujihashi, S. Influence of shot peening condition on the fatigue strength of the carburizing steel. *Electr. Furn. Steel* **2008**, *79*, 69–76. [CrossRef]

54. Koshimune, M.; Matsui, K.; Takahashi, K.; Nakano, W.; Ando, K. Influence of hardness and residual stress on fatigue limit for high strength steel. *Trans. Jpn. Soc. Spring Eng.* **2009**, *54*, 19–26. [CrossRef]

55. Matsui, K.; Koshimune, M.; Takahashi, K.; Ando, K. Influence of shot peening method on rotating bending fatigue limit for high strength steel. *Trans. Jpn. Soc. Spring Eng.* **2010**, *55*, 7–12. [CrossRef]

56. Davies, D.P.; Jenkins, S.L. Influence of stress and environment on the fatigue strength and failure characteristics of case carburized low alloy steels for aerospace applications. *Int. J. Fatigue* **2012**, *44*, 234–244. [CrossRef]

57. Bagherifard, S.; Guagliano, M. Fatigue behavior of a low-alloy steel with nanostructured surface obtained by severe shot peening. *Eng. Fract. Mech.* **2012**, *81*, 56–68. [CrossRef]

58. Dalaei, K.; Karlsson, B. Influence of shot peening on fatigue durability of normalized steel subjected to variable amplitude loading. *Int. J. Fatigue* **2012**, *38*, 75–83. [CrossRef]

59. Bagherifard, S.; Fernandez-Pariente, I.; Ghelichi, R.; Guagliano, M. Fatigue behavior of notched steel specimens with nanaocrystallized surface obtained by severe shot peening. *Mater. Des.* **2013**, *46*, 497–503. [CrossRef]

60. Natori, M.; Song, S.; Sugimoto, K. The effects of fine particle peening on surface residual stress of a TRIP-aided bainitic ferrite steel. *J. Soc. Mater. Sci. Jpn.* **2014**, *63*, 662–668. [CrossRef]

61. Natori, M.; Mizuno, Y.; Song, S.; Sugimoto, K. Effects of fine particle peening on fatigue strength of TRIP-aided bainitic ferrite steel. *J. Soc. Mater. Sci. Jpn.* **2015**, *64*, 620–627. [CrossRef]

62. Sugimoto, K.; Mizuno, Y.; Natori, M.; Hojo, T. Effects of fine particle peening on fatigue strength of a TRIP-aided martensitic steel. *Int. J. Fatigue* **2017**, *100*, 206–214. [CrossRef]

63. Sugimoto, K.; Hojo, T.; Mizuno, Y. Surface-hardened layer properties of newly developed case-hardening steel. *ISIJ Int.* **2018**, *58*, 727–733. [CrossRef]

64. Sugimoto, K.; Hojo, T.; Mizuno, Y. Effects of fine particle peening conditions on the rotational bending fatigue strength of a vacuum-carburized transformation-induced plasticity-aided martensitic steel. *Metall. Mater. Trans. A* **2018**, *49*, 1552–1560. [CrossRef]

65. Sugimoto, K.; Hojo, T.; Mizuno, Y. Effects of vacuum-carburizing conditions on the fatigue strength of a transformation-induced plasticity-aided martensitic steel. *Mater. Sci. Technol.* **2018**, *34*, 743–750. [CrossRef]

66. Koistinen, D.P.; Marburger, R.E. A general equation describing the extent of the austenite-martensite transformation in pure iron carbon alloys and plain carbon steel. *Acta Metall.* **1959**, *7*, 59–60. [CrossRef]

67. Dieter, G.E. *Mechanical Metallurgy (SI Metric Edition)*; McGraw-Hill: Singapore, 1988; p. 403.

68. Knott, J.F. *Fundamentals of Fracture Mechanics*; Baifukan: Tokyo, Japan, 1977; p. 138.

69. Umemoto, M. Nanocrystallization of steels by severe plastic deformation. *Mater. Trans.* **2003**, *44*, 1900–1911. [CrossRef]

70. Sakaki, T.; Sugimoto, K.; Fukuzato, T. Role of internal stress for continuous yielding of dual-phase steels. *Acta Metall.* **1983**, *31*, 1737–1746. [CrossRef]

71. Sugimoto, K.; Sato, S.; Arai, G. Hot forging of ultra high-strength TRIP-aided steel. *Mater. Sci. Forum* **2010**, *638–642*, 3074–3079. [CrossRef]

72. Sugimoto, K.; Kobayashi, J.; Arai, G. Development of low alloy TRIP-aided steel for hot-forging parts with excellent toughness. *Steel Res. Int.* **2010**, *81*, 254–257.

73. Bleck, W.; Prahl, U.; Hirt, G.; Bambach, M. Designing new forging steels by ICMPE. In *Advanced in Production Technology*; Brecher, C., Ed.; Springer International Publishing AG: Cham, Switzerland, 2015; pp. 85–98.

Article

Effects of Alloying Elements Addition on Delayed Fracture Properties of Ultra High-Strength TRIP-Aided Martensitic Steels

Tomohiko Hojo [1,*], Junya Kobayashi [2], Koh-ichi Sugimoto [3], Akihiko Nagasaka [4] and Eiji Akiyama [1]

[1]   Institute for Materials Research, Tohoku University, Sendai 980-8577, Japan; akiyama@imr.tohoku.ac.jp
[2]   Department of Mechanical Systems Engineering, College of Engineering, Ibaraki University, Ibaraki 316-8511, Japan; junya.kobayashi.jkoba@vc.ibaraki.ac.jp
[3]   Department of Mechanical Systems Engineering, School of Science and Technology, Shinshu University, Nagano 380-8553, Japan; sugimot@shinshu-u.ac.jp
[4]   Department of Mechanical Engineering, National Institute of Technology, Nagano College, Nagano 381-8550, Japan; nagasaka@nagano-nct.ac.jp
*   Correspondence: hojo@imr.tohoku.ac.jp; Tel.: +81-22-215-2062

Received: 20 November 2019; Accepted: 17 December 2019; Published: 19 December 2019

Abstract: To develop ultra high-strength cold stamping steels for automobile frame parts, the effects of alloying elements on hydrogen embrittlement properties of ultra high-strength low alloy transformation induced plasticity (TRIP)-aided steels with a martensite matrix (TM steels) were investigated using the four-point bending test and conventional strain rate tensile test (CSRT). Hydrogen embrittlement properties of the TM steels were improved by the alloying addition. Particularly, 1.0 mass% chromium added TM steel indicated excellent hydrogen embrittlement resistance. This effect was attributed to (1) the decrease in the diffusible hydrogen concentration at the uniform and fine prior austenite grain and packet, block, and lath boundaries; (2) the suppression of hydrogen trapping at martensite matrix/cementite interfaces owing to the suppression of precipitation of cementite at the coarse martensite lath matrix; and (3) the suppression of the hydrogen diffusion to the crack initiation sites owing to the high stability of retained austenite because of the existence of retained austenite in a large amount of the martensite–austenite constituent (M–A) phase in the TM steels containing 1.0 mass% chromium.

Keywords: TRIP-aided steels; microalloying; hydrogen embrittlement; retained austenite

---

1. Introduction

Recently, ultra high-strength steels with a tensile strength of more than 980 MPa have been applied for automobile frame parts to improve the impact safety and fuel efficiency of the vehicles. In the ultra high-strength steels, 980–1180 MPa grade transformation induced plasticity (TRIP)-aided bainitic ferrite steels (TBF steels) associated with transformation induced plasticity (TRIP) [1] of retained austenite have been used for the ultra high-strength automobile structural parts because of the excellent press formability [2], impact properties [3], fatigue properties [4] and hydrogen embrittlement (delayed fracture) resistance [5]. In addition, 1500 MPa grade hot stamping ultra high-strength steels [6] have been applied for the part of automobile impact safety components such as center pillars and side impact bars. Generally, it is known that hydrogen embrittlement remarkably appears in the structural high strength bolts when the tensile strength of steels increases to more than 980 MPa. Thus, occurrence of the hydrogen embrittlement becomes a serious problem in the automobile ultra high-strength steel sheets.

Among the 1470 MPa grade ultra high-strength steels, the ultra high-strength TRIP-aided martensitic steels (TM steels) are expected to be applied for automobile structural parts, as the TM steels exhibit ultra high-strength of 1500 MPa grade owing to the martensite lath matrix and excellent formability because of TRIP effects of retained austenite existing in the martensite–austenite constituent (M–A) phase. However, further strengthening of the TM steels requires improvement of hardenability and tailoring the uniform microstructure. In this study, we produced the TM steels microalloyed with elements such as chromium, molybdenum, nickel, titanium, and boron and investigated the hydrogen embrittlement properties of steels using a constant load test and a conventional strain rate tensile test (CSRT) [7–9], aiming at applying the steels to the automotive structural parts.

2. Materials and Methods

In this study, six kinds of cold rolled 0.2C–1.5Si–1.5Mn (mass%) steel sheets with a thickness of 1.2 mm with microalloying elements of niobium, chromium, molybdenum, nickel, boron, and titanium were prepared. The chemical compositions of these steels are listed in Table 1. The martensite transformation start temperature (M_S) [10] calculated by the following equation is also listed in the Table.

$$M_S \, (^\circ\mathrm{C}) = 550 - 361 \times (\%\mathrm{C}) - 39 \times (\%\mathrm{Mn}) - 0 \times (\%\mathrm{Si}) + 30 \times (\%\mathrm{Al}) - 5 \times (\%\mathrm{Mo}), \tag{1}$$

where %C, %Mn, %Si, %Al, and %Mo in mass% are alloying contents in the steels, respectively. Moreover, hardenability of steels defined as a product of the multiplying factors (Πfi) [11], estimated by the following equation, is also shown in Table 1.

$$\Pi\mathrm{fi} = (1 + 0.64\%\mathrm{Si}) \times (1 + 4.10\%\mathrm{Mn}) \times (1 + 2.83\%\mathrm{P}) \times (1 - 0.62\%\mathrm{S}) \times (1 + 2.33\%\mathrm{Cr}) \times$$
$$(1 + 0.52\%\mathrm{Ni}) \times (1 + 3.14\%\mathrm{Mo}) \times (1 - 0.27\%\mathrm{Cu}) \times (1 + 1.5(0.9 - \%\mathrm{C})), \tag{2}$$

where (1 + 1.5 (0.9 − %C)) was added when B was added.

The TRIP-aided steels with martensitic matrix were produced by annealing at 920 °C × 1200 s followed by partitioning treatment (T_p) at 250 °C for steels A and B and 350 °C for steels C, D, E, and F for 1000 s, as shown in Figure 1.

Retained austenite characteristics were analyzed by X-ray diffractometry (Rigaku Co. Ltd., RINT2000, Tokyo, Japan). Volume fraction of retained austenite (f_γ (vol.%)) was calculated from integrated intensity of 200_α, 211_α, 200_γ, 220_γ, and 311_γ diffraction peaks measured using CuKα radiation [12]. On the other hand, carbon concentration of retained austenite (C_γ (mass%)) was estimated by the following equation (3) [13] using an average lattice parameter (a_γ ($\times10^{-10}$ m)) evaluated from 200_γ, 220_γ, and 311_γ peaks analyzed using CuKα radiation.

$$a_\gamma = 3.5780 + 0.0330C_\gamma + 0.00095Mn_\gamma - 0.0002Ni_\gamma + 0.0006Cr_\gamma + 0.0056Al_\gamma + 0.0220N_\gamma +$$
$$0.0004Co_\gamma + 0.0015Cu_\gamma + 0.0051Nb_\gamma + 0.0031Mo_\gamma + 0.0039Ti_\gamma + 0.0018V_\gamma + 0.0018W_\gamma, \tag{3}$$

where Mn_γ, Ni_γ, Cr_γ, Al_γ, N_γ, Co_γ, Cu_γ, Nb_γ, Mo_γ, Ti_γ, V_γ, and W_γ represent alloying contents in austenite. In this study, the average content of each element in the steels was used.

The four-point bending cathode charging technique [5,8,14] was adopted to evaluate the hydrogen embrittlement properties of the steels using rectangular specimens of 65 mm in length, 10 mm in width, and 1.2 mm in thickness, with hydrogen charging in a 0.5 mol H_2SO_4 (FUJIFILM Wako Pure Chemical Co. Ltd., Sulfuric Acid, Japan) + 0.01 mol NH_4SCN (FUJIFILM Wako Pure Chemical Co. Ltd., Ammonium Thiocyanate, Japan) solution at 25 °C at a current density of 100 A/m^2. Hydrogen embrittlement properties were evaluated by delayed fracture strength (*DFS*), which is the maximum bending stress that does not cause failure of the specimen for 5 h.

Tensile tests of the steels were also carried out using a tensile testing machine (Shimadzu Co. Ltd., AG-X plus 100kN, Kyoto, Japan) at a crosshead speed of 1 mm/min at 25 °C without and with hydrogen charging. Tensile testing specimens were used with 15 mm in gauge length, 6 mm in width, and 1.2 mm in thickness. Hydrogen embrittlement properties were evaluated by hydrogen embrittlement susceptibility (*HES*) calculated by the following equation.

$$HES = (1 - \varepsilon_1/\varepsilon_0) \times 100\%,\qquad(4)$$

where ε_0 and ε_1 represent total elongation without and with hydrogen charging, respectively. Hydrogen charging to the tensile specimens was conducted by cathodic charging. A hydrogen charging solution of a 3% NaCl (FUJIFILM Wako Pure Chemical Co. Ltd., Sodium Chloride, Japan) + 3 g/L NH_4SCN was used, and the hydrogen charging condition was at a current density of 10 A/m^2, at 25 °C for 48 h.

Diffusible hydrogen concentration of the tensile specimens was measured by means of thermal desorption analysis (TDA) using a gas chromatography as a hydrogen detector (J-SCIENCE LAB Co. Ltd., GC-MS, Kyoto, Japan). Specimens were heated from room temperature to 800 °C at a heating rate of 200 °C/h. Diffusible hydrogen concentration (wt. ppm) was defined as a weight ppm of evolved hydrogen from room temperature to 150 °C corresponding to a first peak of the hydrogen evolution curve to specimen weight. The hydrogen charged specimens were kept in liquid nitrogen to prevent the hydrogen evolution from the specimen until hydrogen analysis.

Electron Back Scatter Diffraction Patterns (EBSD) analysis for the body centered cubic (bcc) phase was carried out in an area of 40 μm × 40 μm with step size of 0.2 μm. Inverse pole figure (IPF) and kernel average misorientation (KAM) maps were used to define the initial microstructures and the strain distributions after the tensile test in this study. Moreover, image quality (IQ) value was used to estimate an area fraction of the M–A phase. The area fraction of the M–A phase was defined as a total fraction of IQ value below 50.

In this study, the four-point bending constant load test and conventional strain rate tensile test were conducted at different hydrogen charging conditions. Four-point bending tests were conducted at relatively low stress with a high hydrogen concentration, whereas stress and plastic strain were continuously increased during conventional strain rate tensile tests with a low hydrogen concentration. These two testing and hydrogen charging methods were adopted to evaluate the hydrogen embrittlement properties of the microalloyed TM steels at a wide range of conditions of stress, plastic strain, and hydrogen concentration.

Table 1. Chemical compositions (mass%), martensite transformation start temperature (M_S) and hardenability (Πfi) of transformation induced plasticity (TRIP)-aided steels with a martensite matrix (TM steels).

Steel	C	Si	Mn	Ni	Cr	Mo	Al	Nb	Ti	B	M_S	Πfi
A	0.20	1.50	1.50	-	-	-	0.039	-	-	-	420	14.60
B	0.20	1.52	1.50	-	-	-	0.039	0.05	0.02	0.0018	420	29.21
C	0.21	1.49	1.50	-	0.50	-	0.04	0.05	-	-	407	30.55
D	0.20	1.49	1.50	-	1.00	-	0.04	0.05	-	-	401	46.99
E	0.18	1.48	1.49	-	1.02	0.20	0.043	0.05	-	-	407	76.82
F	0.21	1.49	1.49	1.52	1.00	0.20	0.034	0.049	-	-	370	135.8

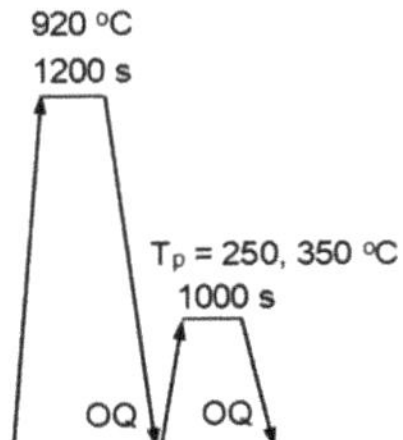

Figure 1. Heat treatment diagram of transformation induced plasticity (TRIP)-aided steels with a martensite matrix (TM steels). OQ represents oil quenching.

3. Results

3.1. Microstructure and Tensile Properties

Figure 2 shows inverse pole figure (IPF) maps of the TM steels analyzed by EBSD. Typical transmission electron micrographs of the TM steels are shown in Figure 3. The martensite lath matrix of the TM steels became fine and uniform owing to the addition of alloying elements in comparison with that of steel A (Figures 2 and 3). The TM steels consisted of wide lath martensite, a hard and narrow M–A phase, and retained austenite (Figure 3). Retained austenite film existed at lath boundaries and the M–A phase consisted of narrow martensite and retained austenite in steel A. In contrast, retained austenite film was not observed at martensite lath boundaries in steels B–F, although fine retained austenite existed in the M–A phase. Precipitation of cementite was suppressed by the alloying addition in the TM steels, although it was confirmed that cementite precipitated at coarse lath martensite in steel A without alloying addition (Figure 3).

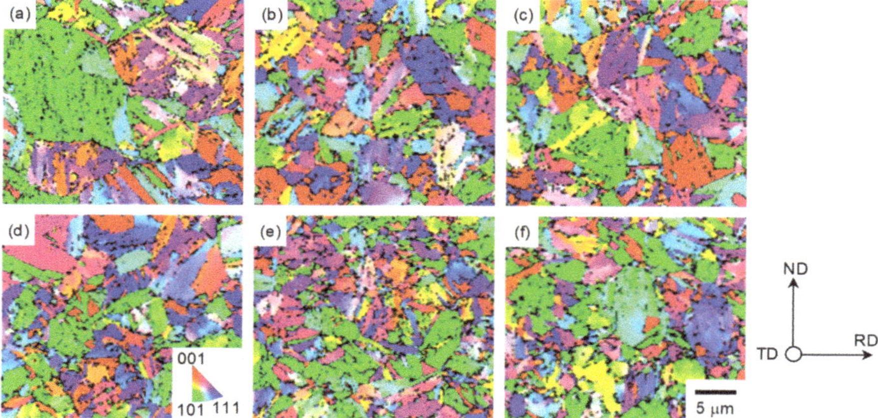

Figure 2. Microstructures of TM steels of steels (**a**) A, (**b**) B, (**c**) C, (**d**) D, (**e**) E, and (**f**) F partitioned at 250 or 350 °C. RD, ND and TD represent rolling, normal, and transverse directions, respectively.

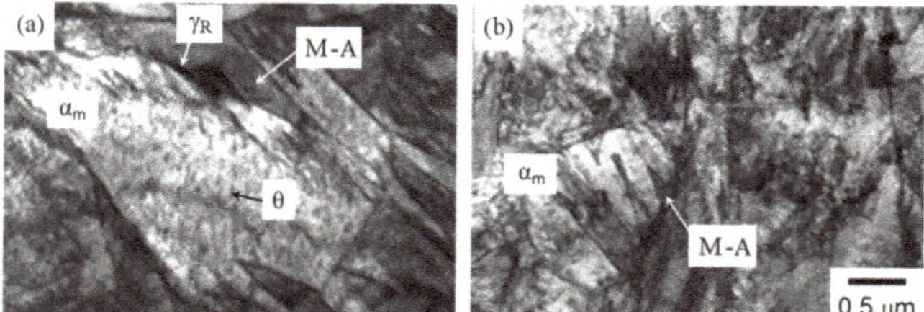

Figure 3. Transmission electron micrographs of TM steels of steels (**a**) A and (**b**) D partitioned at (**a**) 250 and (**b**) 350 °C, in which γ_R, α_m, θ, and M–A represent retained austenite, martensite, cementite, and martensite–austenite constituent (M–A) phase, respectively.

Table 2 shows the retained austenite characteristics, area fraction of the M–A phase, tensile properties, and prior austenite grain size of the TM steels. The initial volume fractions of retained austenite ($f_{\gamma 0}$s) were between 1.8 vol.% and 2.2 vol.%, and its carbon concentrations ($C_{\gamma 0}$s) were 0.52 mass%–1.32 mass%. $f_{\gamma 0}$s were slightly changed by the alloying addition. On the other hand, chromium addition increased $C_{\gamma 0}$ and complex addition of titanium–boron, chromium–molybdenum, and nickel–chromium–molybdenum significantly decreased $C_{\gamma 0}$ in the TM steels compared with steel

A. The area fraction of M–A phases ($f_{\text{M–A}}$s) of the TM steels was 20.0% or more. Alloying addition increased $f_{\text{M–A}}$ in the TM steels.

Table 2. Tensile properties, retained austenite characteristics, area fraction of the martensite–austenite constituent (M–A) phase, and prior austenite grain size of TM steels.

Steel	*TS*	*YS*	*TEl*	*UEl*	$f_{\gamma 0}$	$C_{\gamma 0}$	$f_{\text{M–A}}$	*d*
A	1422	1137	12.6	4.2	2.1	1.20	20.0	22.6
B	1490	1292	13.9	5.6	2.0	1.32	21.4	16.0
C	1445	1331	12.6	4.7	1.8	0.75	22.3	14.3
D	1498	1386	14.2	5.1	2.2	1.19	20.7	14.8
E	1496	1376	15.3	5.1	2.2	0.52	24.9	11.3
F	1547	1396	14.1	4.3	1.9	0.96	26.7	12.7

TS (MPa): tensile strength, *YS* (MPa): yield strength, *TEl* (%): total elongation, *UEl* (%): uniform elongation, $f_{\gamma 0}$ (vol.%): initial volume fraction of retained austenite, $C_{\gamma 0}$ (mass%): initial carbon concentration of retained austenite, $f_{\text{M–A}}$ (%): area fraction of M–A phase, *d* (µm): prior austenite grain diameter.

Tensile strengths (*TS*s) of the TM steels were between 1422 and 1542 MPa, and total elongations (*TEl*s) exhibited 12.6–15.3%. Alloying addition to the TM steels resulted in both high strength and large elongation compared with steel A, and they exhibited an excellent strength–ductility balance. Particularly, the tensile strength (*TS*) increased owing to the Ni–Cr–Mo complex addition, and the total elongation (*TEl*) was increased by the complex addition of Cr–Mo.

The prior austenite grain size (*d*) of steel A was 22.6 µm, and alloyed TM steel exhibited an obviously decreased prior austenite grain size in the range of 11.3–16.0 µm.

3.2. Hydrogen Enbrittlement Properties Evaluated by Four-Point Bending Test

Typical applied stress-fracture time plots of the TM steels are shown in Figures 4 and 5 shows relationship between delayed fracture strength (*DFS*), which is the maximum bending stress causing no failure of specimen for 5 h and tensile strength (*TS*). In the all steels, time to fracture (t_f) was shortened with applied stress (σ_A). Steel C alloyed with 0.5 mass% chromium possessed a long time to failure and increased *DFS* in comparison with other TM steels. Although steel D alloyed with 1.0 mass% chromium fractured earlier at a high applied bending stress region than steel A, *DFS* of steel D was higher than that of steel A. On the other hand, steel E co-alloyed with chromium and molybdenum fractured in short time compared with the steel A, and *DFS* was similar to that of steel A. From Figure 5, *DFS* of about 750–850 MPa was achieved in steel B alloyed with titanium and boron and C with 0.5 mass% chromium. All of the steels with alloying addition except for steel E showed higher *DFS* than steel A, though the *TS* of the alloyed steels was higher than that of steel A. Steel E showed slightly low *DFS* compared with steel A, but its *TS* was higher than that of steel A. The relationship between *DFS* and *TS* of the alloyed steels in the band is shown by the hatch in the figure, and the relative position of the band to the data point of steel A indicated that the alloying additions had a positive effect on improving the resistance of the TM steel against hydrogen embrittlement.

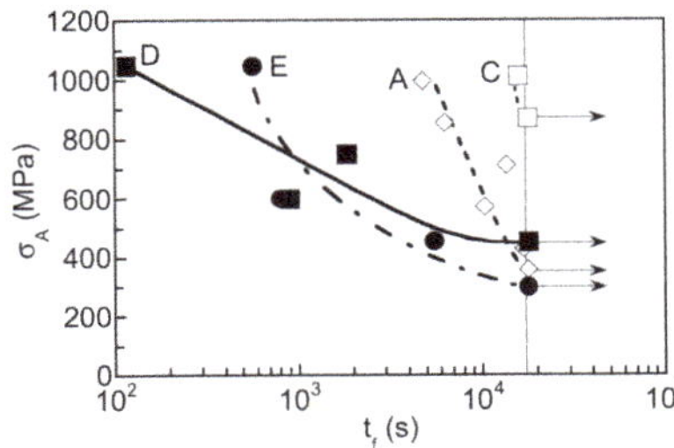

Figure 4. Typical delayed fracture curves of TM steels of steels A, C, D, and E.

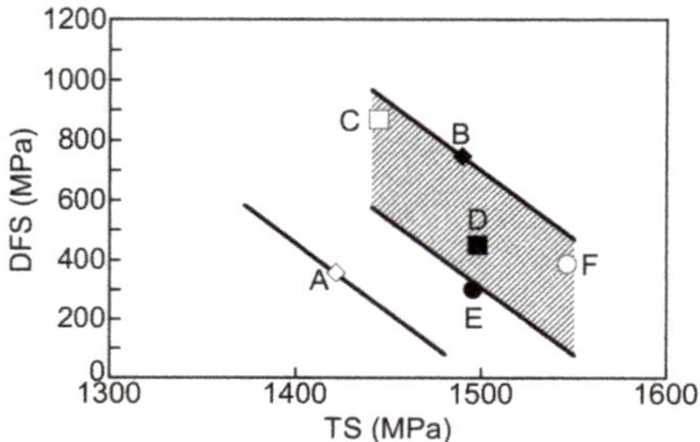

Figure 5. Variations in delayed fracture strength (*DFS*) as a function of tensile strength (*TS*) in TM steels.

Figure 6 shows the fracture surface of the TM steels after four-point bending testing. In Figure 6, four-point bending tests were carried out at an applied bending stress of 70–80 MPa higher than the value of *DFS*. Fracture surfaces of the steels indicated a mixed fracture of intergranular and transgranular fractures. The ratio of intergranular fracture surface was high for steels A and E, suggesting the correspondence of the morphologies with the susceptibilities evaluated by *DFS*. It was noted that a small unit crack path was observed in the alloyed TM steels compared with steel A, presumably owing to the small size of prior austenite grain, as shown in Table 2. The morphology of the fracture surface of steel B was similar to that of other microalloyed TM steels.

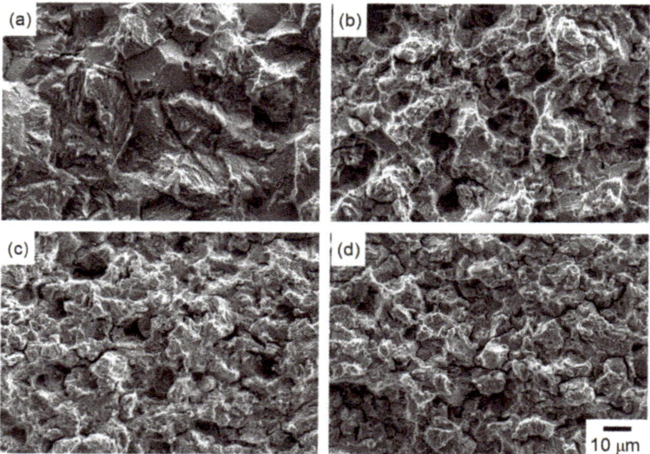

Figure 6. Typical fracture surfaces of steels (**a**) A, (**b**) C, (**c**) D, and (**d**) E after four-point bending testing.

3.3. Hydrogen Embrittlement Properties Evaluated by Tensile Test

Figure 7 shows typical stress–strain curves without and with hydrogen charging in the TM steels. Steels A and E charged with hydrogen fractured in an early stage with a deterioration of local elongation, although the stress–strain behavior up to a maximum stress with hydrogen was almost the same as that without hydrogen. A decrease in the local elongation hardly appeared in steels C and D with additive chromium of 0.5 mass% and 1.0 mass%, respectively. The relationships between hydrogen embrittlement susceptibility (*HES*) and yield strength (*YS*) and tensile strength (*TS*) are shown in Figure 8. Here, the low *HES* stands for excellent hydrogen embrittlement resistance; that is, *HES* = 0% means that hydrogen embrittlement never occurs. The hydrogen embrittlement susceptibility decreased owing to the addition of alloying elements in the TM steels, although yield and tensile strengths increased. Particularly, steel D alloyed with 1.0 mass% chromium exhibited low *HES*. Figure 9 shows fracture surfaces of steels A, C, D, and E without and with hydrogen charging. In all of the TM steels, dimple fracture occurred when tensile tests were conducted without hydrogen charging. The size of the dimple was quite small in steels A, C, and D, and fine and coarse dimple coexisted in steel E. On the other hand, steel A with hydrogen charging indicated a large size of dimple, whereas the fracture surface of the dimple decreased and that of quasi cleavage increased by the hydrogen charging in steel C. Moreover, steel D with hydrogen charging showed fine dimples, and steel E with hydrogen exhibited coarse dimple fracture, as well as the fracture surfaces of these steels without hydrogen charging. Hydrogen evolution curves of the TM steels charged with hydrogen are shown in Figure 10. It was confirmed that a large amount of hydrogen was evolved from room temperature to around 100 °C in the TM steels. The concentrations of diffusible hydrogen that was evolved at temperature between room temperature and 150 °C in the hydrogen evolution curves are listed in Table 3. The diffusible hydrogen concentration was increased by the addition of alloying element.

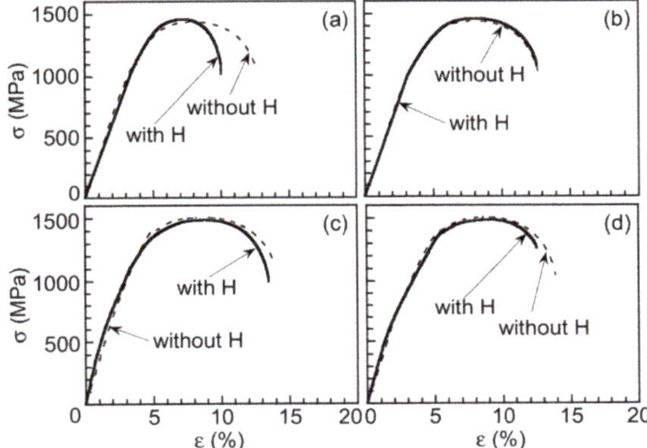

Figure 7. Typical stress–strain curves of steels (**a**) A, (**b**) C, (**c**) D, and (**d**) E without and with diffusible hydrogen.

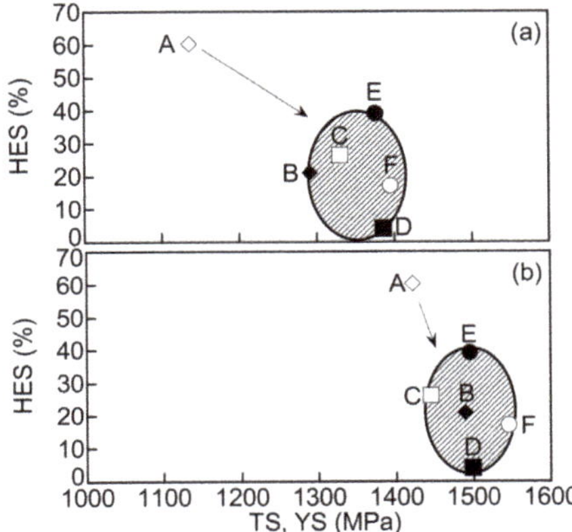

Figure 8. Relationships between hydrogen embrittlement susceptibility (*HES*) and (**a**) yield strength (*YS*) and (**b**) tensile strength (*TS*) in TM steels.

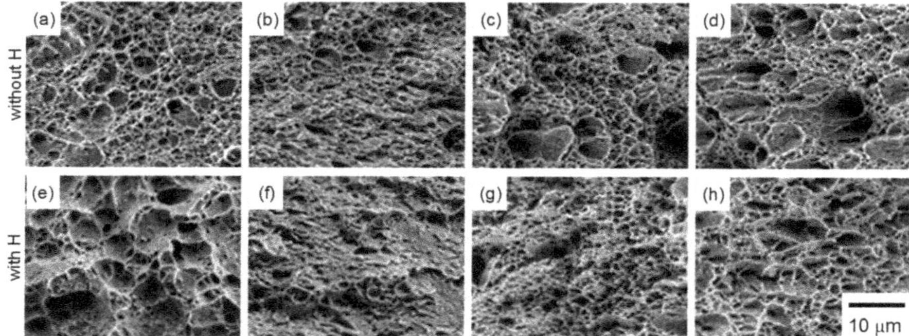

Figure 9. Scanning electron micrographs of the fracture surface tested by tensile testing of steels (**a**,**e**) A, (**b**,**f**) C, (**c**,**g**) D, and (**d**,**h**) E (**a**–**d**) without and (**e**–**h**) with hydrogen.

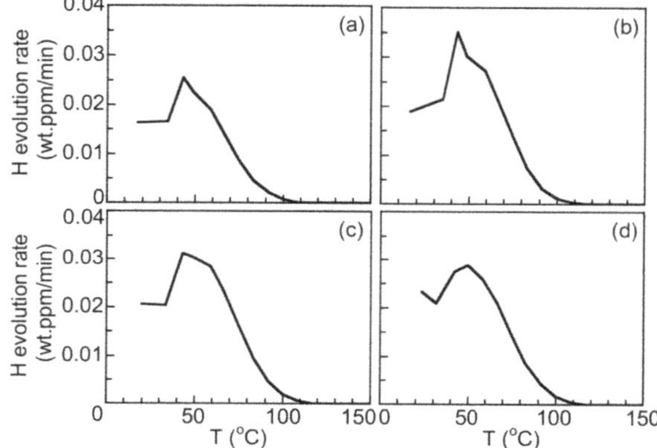

Figure 10. Hydrogen evolution curves of steels (**a**) A, (**b**) C, (**c**) D, and (**d**) E.

Table 3. Diffusible hydrogen concentration (H_D) of TM steels.

Steels	H_D (wt. ppm)
A	0.65
B	0.85
C	0.90
D	0.93
E	0.88
F	0.96

4. Discussion

Generally, the hydrogen embrittlement properties of the high-strength low alloy TRIP-aided steels, such as TBF steels containing a large amount of retained austenite, were affected by microstructural sizes, retained austenite characteristics, residual stress, and plastic strain distributions owing to the press forming and hydrogen concentration. Thus, we are going to discuss the relationship between hydrogen embrittlement properties and those dominant factors in the following parts.

4.1. Improvement of Hydrogen Embrittlement Properties by Refinement of Microstructure

It is known that the hydrogen embrittlement resistance of high strength steels depends on the size of a microstructure, namely, refinement of grain, packet, block, and lath sizes reduces the risk of void and crack initiation, and improves the hydrogen embrittlement resistance because of the decrease in the facet size [15]. Moreover, it was suggested that hydrogen trapped at matrix/cementite interfaces decreased the hydrogen embrittlement resistance owing to the diffusion of hydrogen to the crack initiation sites such as the prior austenite grain and lath boundaries, as these interfaces are weak hydrogen trapping sites [16]. On the other hand, many investigations of the improvement of the hydrogen embrittlement resistance of high-strength steels have been strongly conducted with microalloying such as vanadium [17] and titanium [17,18], because matrix/metallic carbide coherency strain fields were the effective hydrogen trapping sites for the hydrogen embrittlement resistance due to the high hydrogen trapping energy.

The TM steels (steels B–E) in this study exhibited a small size of prior austenite grain, packet, block, and martensite lath because of the alloying addition (Figure 2). In addition, precipitation of cementite in the martensite laths and at the prior austenite grain boundaries was suppressed in the TM steel with 1.0 mass% chromium (steel D) (Figure 3). The refinement of prior austenite grain size owing to the addition of alloying element might be attributed to the precipitation of fine NbC during the hot rolling process, which suppressed growth of prior austenite grain [19]. Moreover, the improved hardenability of the microalloyed TM steels suppressed the precipitation of cementite in the martensite laths. It is also considered that the decrease in the M_S temperature delays the martensite transformation at quenching, and increases the amount of the M–A phase. The mechanism of such a microstructural change of the TM steels owing to the addition of alloying elements has been reported in previous reports [12,20,21].

It was considered that the microstructural change as mentioned above in the TM steels containing alloying elements improved hydrogen embrittlement resistance, namely, the hydrogen embrittlement resistance of the TM steels was improved by the suppression of the void and the crack initiation and propagation at the prior austenite grain, packet, block, and lath boundaries owing to the refinement of these hydrogen trapping sites, by the suppression of hydrogen trapping at martensite/cementite interfaces because of the no precipitation of cementite and because of the increase in the amount of the M–A phase. The effects of the M–A phase on the hydrogen embrittlement resistance are discussed later.

4.2. Effects of Stability of Retained Austenite on Hydrogen Embrittlement Properties

The hydrogen embrittlement properties were greatly affected by the retained austenite characteristics. In this study, the initial carbon concentration affected the hydrogen embrittlement resistance, although the initial volume fraction of retained austenite hardly affected *DFS* and *HES*, as shown in Figure 11. It was considered that transformation of unstable retained austenite to martensite was accelerated when unstable retained austenite absorbed hydrogen [5]. Moreover, the crack and void initiation was promoted by the desorption of hydrogen from transformed martensite because hydrogen solubility is significantly decreased when retained austenite (face centered cubic (fcc) phase) transforms to martensite (bcc phase) [22]. Hojo et al. [5] have investigated the hydrogen embrittlement resistance of aluminum and manganese added TRIP-aided bainitic ferrite (TBF) steels, and have reported that aluminum addition improved the hydrogen embrittlement resistance, whereas manganese addition worsened it. This was attributed to the increase in the stability of retained austenite in the aluminum added TBF steels, whereas the manganese addition to the TBF steels deteriorated the stability of retained austenite. The initial carbon concentration of steel D with additive chromium was similar to that in steel A, although the initial carbon concentration of steels B, E, and F was low compared with that of steel A. Thus, it was suggested that the chromium added TM steels exhibited excellent hydrogen embrittlement resistance owing to the suppression of martensite transformation during hydrogen embrittlement tests because of the high stability of retained austenite.

It was confirmed that the M–A phase formed at triple junctions of prior austenite grain in the TM steels, and retained austenite existed along the martensite lath boundaries in steel A (Figure 3a). However, in the TM steels alloyed with additive elements, retained austenite tended not to exist alone, but to coexist in the M–A phase. In addition, it was confirmed that the area fraction of the

M–A phase measured by image quality of EBSD increased owing to the alloying addition, as shown in Table 2. The filmy retained austenite in steel A might easily transform to martensite, assisted by straining or hydrogen despite the high initial carbon concentration of retained austenite, because retained austenite existed along relatively large and soft lath martensite, which formed during rapid cooling between austenitizing and ambient temperatures. On the other hand, it was considered that the stability of retained austenite in the TM steels alloyed with additive elements was high because fine retained austenite in the M–A phase was surrounded by the fine and hard martensite and, consequently, the martensite transformation of retained austenite in the M–A phase was suppressed [12,20,23]. Therefore, the TM steels with alloying addition exhibited excellent hydrogen embrittlement resistance.

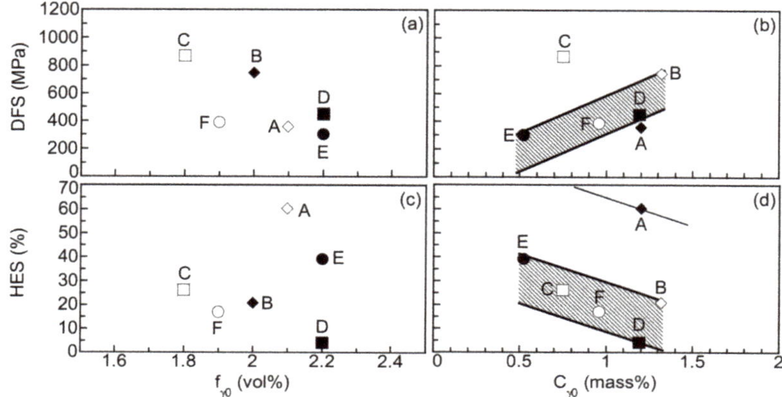

Figure 11. Variations in (**a**,**b**) delayed fracture strength (*DFS*) and (**c**,**d**) hydrogen embrittlement susceptibility (*HES*) as functions of (**a**,**c**) initial volume fraction of retained austenite ($f_{\gamma 0}$) and (**b**,**d**) its carbon concentration ($C_{\gamma 0}$) in TM steels.

4.3. Hydrogen Absorption Properties and Hydrogen Embrittlement Resistance

It is known that the amount of diffusible hydrogen and its trapping site in steels significantly affect the hydrogen embrittlement resistance of steels [24,25]. As shown in Figure 10 and Table 3, it was confirmed that a large amount of hydrogen was evolved from room temperature to around 100 °C in the TM steels. Furthermore, the addition of alloying element increased the amount of absorbed hydrogen. It is known that the hydrogen evolution from this temperature range is associated with hydrogen trapped at prior austenite grain and lath boundaries [26], at dislocations [25], at matrix/carbide interfaces [16], and in retained austenite [27,28] or at matrix/retained austenite interfaces [28]. In the TM steels, it was assumed that the amount of diffusible hydrogen in the TM steels increased because of a large amount of hydrogen trapped at refined prior austenite grain and lath boundaries owing to the addition of alloying element. However, the delayed fracture strength obtained by the four-point bending test was improved and hydrogen embrittlement susceptibility evaluated by the tensile test decreased, although diffusible hydrogen concentration increased compared with steel A, as shown in Figure 12. Presumably, areas of prior austenite grain and lath boundaries increased because prior austenite grain became small, and the width and length of laths decreased. Therefore, it was suggested that the occurrence of hydrogen embrittlement caused by void and crack nucleation, as well as their growth and propagation, were suppressed, resulting from low hydrogen concentration at prior austenite grain and lath boundaries, although the total amount of hydrogen trapped at prior austenite grain and lath boundaries was high owing to the increased area of boundaries.

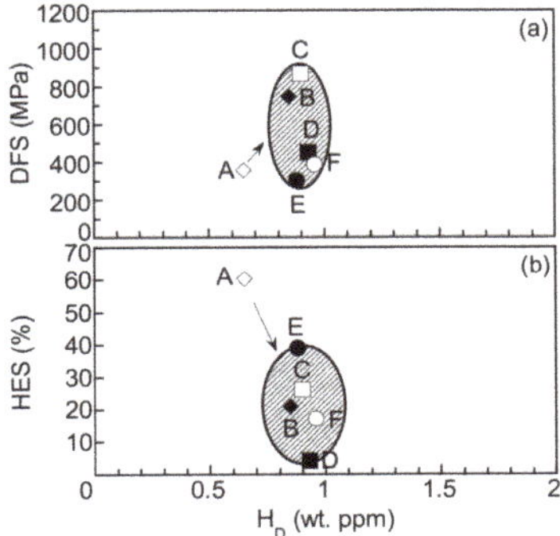

Figure 12. Relationships between (**a**) delayed fracture strength (*DFS*) and (**b**) hydrogen embrittlement susceptibility (*HES*) and diffusible hydrogen concentration (H_D) in TM steels.

4.4. Mechanism of Hydrogen Embrittlement in TM steels

Hydrogen embrittlement occurred with a large amount of plastic deformation given by the tensile test in this study. Accordingly, strain distributions at uniformly deformed part of tensile specimens in steels A and D were compared using kernel average misorientation (KAM) map analyzed by EBSD. The KAM maps and KAM value variations are shown in Figures 13 and 14, respectively. When tensile tests were carried out at the conventional strain rate without hydrogen charging, the KAM values of both steels A and D at both vicinity and inside of the packet and the lath boundaries slightly increased in comparison with those of as-heat treated steels (Figure 13). In addition, the high KAM value uniformly distributed in steel D compared with steel A (Figures 13 and 14). On the other hand, when tensile tests were conducted with hydrogen charging, the KAM value at vicinity of the packet and the lath boundaries of steel A significantly increased, and the distribution of the KAM value of steel D was hardly changed in comparison with that of steel D tensile tested without hydrogen charging, although the tendency of the frequency of the KAM value of steel A was similar to that of steel D (Figure 14). These results suggested that the hydrogen embrittlement remarkably occurred owing to the acceleration of the localized deformation such as the vicinity of the packet and the lath boundaries in steel A, whereas steel D with hydrogen exhibited a similar fracture morphology as steel D without hydrogen because of the suppression of the localized deformation.

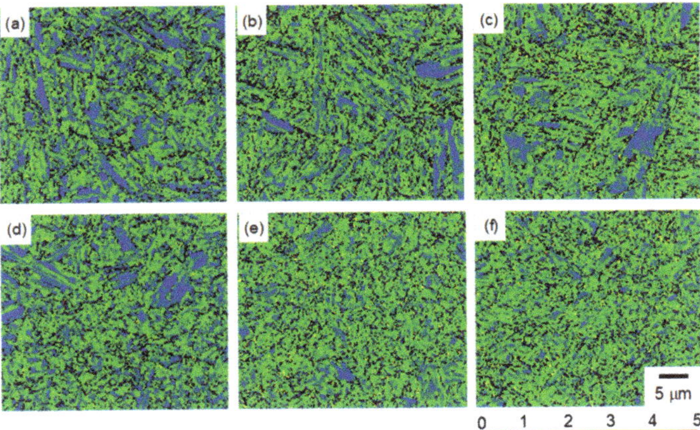

Figure 13. Kernel average misorientation (KAM) maps of (**a,d**) as heat treated, (**b,e**) deformed without hydrogen, and (**c,f**) deformed with hydrogen in steels (**a–c**) A and (**d–f**) D.

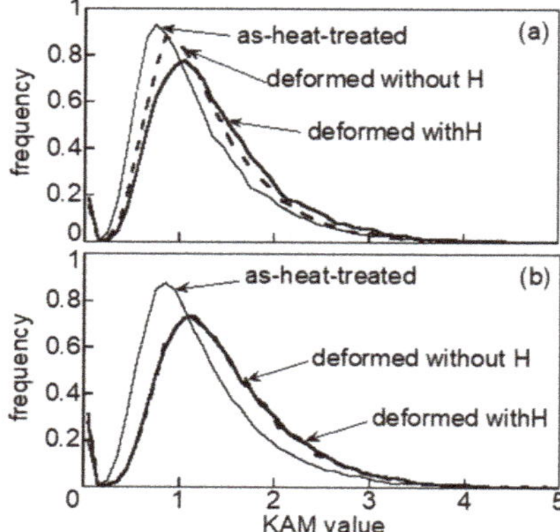

Figure 14. Variations in frequency as a function of kernel average misorientation (KAM) values of as-heat-treated, deformed without hydrogen, and deformed with hydrogen in steels (**a**) A and (**b**) D.

5. Conclusions

To apply the TRIP-aided martensitic steels for the automobile structural steels as third generation ultra high-strength steels, the effects of microalloying elements added for high hardenability on hydrogen embrittlement properties were investigated. The results are summarized as follows.

(1) Hydrogen embrittlement resistance was improved by the addition of alloying elements, especially 1.0 mass% chromium addition was most effective to improve the hydrogen embrittlement resistance.

(2) Higher hydrogen embrittlement resistance of the microalloyed TRIP-aided martensitic steels was obtained by (i) the suppression of initiation and propagation of voids and cracks at prior austenite grain, packet, block, and lath boundaries owing to the refinement of their sizes; (ii) the suppression of the hydrogen trapping at martensite matrix/cementite interfaces because of the hindered precipitation of cementite; and (iii) the restriction of martensitic transformation of retained austenite because of the high stability of retained austenite.

Author Contributions: Conceptualization, T.H., K.S., and A.N.; methodology, T.H.; formal analysis, T.H. and J.K.; investigation, T.H., J.K., and A.N.; resources, T.H. and E.A.; data curation, T.H.; writing—original draft preparation, T.H.; writing—review and editing, K.S. and E.A.; visualization, T.H.; project administration, T.H. All authors have read and agreed to the published version of the manuscript.

Funding: This work was conducted with financial support by the JKA, Suzuki Foundation, and Okayama Foundation for Science and Technology.

Conflicts of Interest: The authors declare no conflict of interest. The funders had no role in the design of the study; in the collection, analyses, or interpretation of data; in the writing of the manuscript; or in the decision to publish the results.

References

1. Zackay, V.F.; Parker, E.R.; Fahr, D.; Bush, R. The enhancement of ductility in high-strength steels. *Trans. Am. Soc. Met.* **1967**, *60*, 252–258.

2. Sugimoto, K.; Nakano, K.; Song, S.M.; Kashima, T. Retained austenite characteristics and stretch-flangeability of high-strength low-alloy TRIP type bainitic sheet steels. *ISIJ Int.* **2002**, *42*, 450–455. [CrossRef]

3. Hojo, T.; Kobayashi, J.; Sugimoto, K. Impact properties of low-alloy transformation-induced plasticity-steels with different matrix. *Mater. Sci. Technol.* **2016**, *32*, 1035–1042. [CrossRef]

4. Song, S.M.; Sugimoto, K.; Kandaka, S.; Futamura, A.; Kobayashi, M.; Masuda, S. Effects of prestraining on high-cycle fatigue strength of high-strength low alloy TRIP-aided steels. *Mater. Sci. Res. Int.* **2003**, *52*, 223–229. [CrossRef]

5.   Hojo, T.; Sugimoto, K.; Mukai, Y.; Ikeda, S. Effects of aluminum on delayed fracture properties of ultra high strength low alloy TRIP-aided steels. *ISIJ Int.* **2008**, *48*, 824–829. [CrossRef]

6.   Zhou, J.; Wang, B.; Huang, M.-d.; Cui, D. Effect of hot stamping parameters on the mechanical properties and microstructure of cold-rolled 22MnB5 steel strips. *Int. J. Miner. Metall. Mater.* **2014**, *21*, 544–555. [CrossRef]

7.   Hagihara, Y. Evaluation of delayed fracture characteristics of high-strength bolt steels by CSRT. *ISIJ Int.* **2012**, *52*, 292–297. [CrossRef]

8.   Takagi, S.; Hagihara, Y.; Hojo, T.; Urushihara, W.; Kawasaki, K. Comparison of hydrogen embrittlement resistance of high strength steel sheets evaluated by several methods. *ISIJ Int.* **2016**, *56*, 685–692. [CrossRef]

9.   Chida, T.; Hagihara, Y.; Akiyama, E.; Iwanaga, K.; Takagi, S.; Hayakawa, M.; Ohishi, H.; Hirakami, D.; Tarui, T. Comparison of constant load, SSRT and CSRT methods for hydrogen embrittlement evaluation using round bar specimens of high strength steels. *ISIJ Int.* **2016**, *56*, 1268–1275. [CrossRef]

10. Tamura, I. *Steel Material Study on the Strength*; Nikkan-Kogyo Shinbun Ltd.: Tokyo, Japan, 1970.

11. Grossmann, M.A. Hardenability calculated from chemical composition. *Trans. AIME* **1942**, *150*, 227–255.

12. Kobayashi, J.; Ina, D.; Nakajima, Y.; Sugimoto, K. Effects of microalloying on the impact toughness of ultrahigh-strength TRIP-aided martensitic steels. *Metall. Mater. Trans. A* **2013**, *44*, 5006–5017. [CrossRef]

13. Dyson, D.J.; Holmes, B. Effect of alloying additions on the lattice parameter of austenite. *J. Iron Steel Inst.* **1970**, *208*, 469–474.

14. Valentini, R.; Tedesco, M.M.; Corsinovi, S.; Bacchi, L.; Villa, M. Investigation of mechanical tests for hydrogen embrittlement in automotive PHS steels. *Metals* **2019**, *9*, 934. [CrossRef]

15. Proctor, R.P.M.; Paxton, H.W. Effect of prior-austenite grain-size on stress-corrosion cracking susceptibility of AISI-4340 steel. *Asm Trans. Q.* **1969**, *62*, 989.

16. Nagao, A.; Hayashi, K.; Oi, K.; Mitao, S. Effect of uniform distribution of fine cementite on hydrogen embrittlement of low carbon martensitic steel plates. *ISIJ Int.* **2012**, *52*, 213–221. [CrossRef]

17. Cho, L.; Seo, E.J.; Sulistiyo, D.H.; Jo, K.R.; Kim, S.W.; Oh, J.K.; Cho, Y.R.; De Cooman, B.C. Influence of vanadium on the hydrogen embrittlement of aluminized ultra-high strength press hardening steel. *Mater. Sci. Eng. A* **2018**, *735*, 448–455. [CrossRef]

18. Takahashi, J.; Kawakami, K.; Kobayashi, Y.; Tarui, T. The first direct observation of hydrogen trapping sites in TiC precipitation-hardening steel through atom probe tomography. *Scr. Mater.* **2010**, *63*, 261–264. [CrossRef]

19. Sugimoto, K.; Murata, M.; Song, S.M. Formability of Al–Nb bearing ultra high-strength TRIP-aided sheet steels with bainitic ferrite and/or martensite matrix. *ISIJ Int.* **2010**, *50*, 162–168. [CrossRef]

20. Van Pham, D.; Kobayashi, J.; Sugimoto, K. Effects of microalloying on stretch-flangeability of ultrahigh-strength TRIP-aided martensitic steel sheets. *ISIJ Int.* **2014**, *54*, 1943–1951. [CrossRef]

21. Kobayashi, J.; Tonegawa, H.; Sugimoto, K. Cold formability of 22SiMnCrB TRIP-aided martensitic sheet steel. *Procedia Eng.* **2014**, *81*, 1336–1341. [CrossRef]

22. Hojo, T.; Koyama, M.; Terao, N.; Tsuzaki, K.; Akiyama, E. Transformation-assisted hydrogen desorption during deformation in steels: Examples of α'- and ε-martensite. *Int. J. Hydrog. Energy* **2019**, *44*, 30472–30477. [CrossRef]

23. Hojo, T.; Kobayashi, J.; Kochi, T.; Sugimoto, K. Effects of thermomechanical processing on microstructure and shear properties of 22SiMnCrMoB TRIP-aided martensitic steel. *Iron Steel Technol.* **2015**, *10*, 102–110.

24. Wang, M.; Akiyama, E.; Tsuzaki, K. Effect of hydrogen and stress concentration on the notch tensile strength of AISI 4135 steel. *Mater. Sci. Eng. A* **2005**, *398*, 37–46. [CrossRef]

25. Bhadeshia, H.K.D.H. Prevention of hydrogen embrittlement in steels. *ISIJ Int.* **2016**, *56*, 24–36. [CrossRef]

26. Momotani, Y.; Shibata, A.; Terada, D.; Tsuji, N. Effect of strain rate on hydrogen embrittlement in low-carbon martensitic steel. *Int. J. Hydrog. Energy* **2017**, *42*, 3371–3379. [CrossRef]

27. Chan, S.L.I.; Lee, H.L.; Yang, J.R. Effect of retained austenite on the hydrogen content and effective diffusivity of martensitic structure. *Metall. Trans. A* **1991**, *22*, 2579–2586. [CrossRef]

28. Gu, J.; Chang, K.D.; Fang, H.; Bai, B. Delayed fracture properties of 1 500 MPa bainite/martensite dual-phase high strength steel and its hydrogen traps. *ISIJ Int.* **2002**, *42*, 1560–1564. [CrossRef]

MDPI

St. Alban-Anlage 66

4052 Basel

Switzerland

Tel. +41 61 683 77 34

Fax +41 61 302 89 18

www.mdpi.com

Metals Editorial Office

E-mail: metals@mdpi.com

www.mdpi.com/journal/metals